BETTER THAN THE REAPER

BETTER
THAN THE
REAPER

PLANO'S HARVESTER HISTORY

Jeanne Valentine

DEDICATION

To the Writers Group of Plano Community Library. You have inspired, supported, and encouraged me through my years of procrastination.

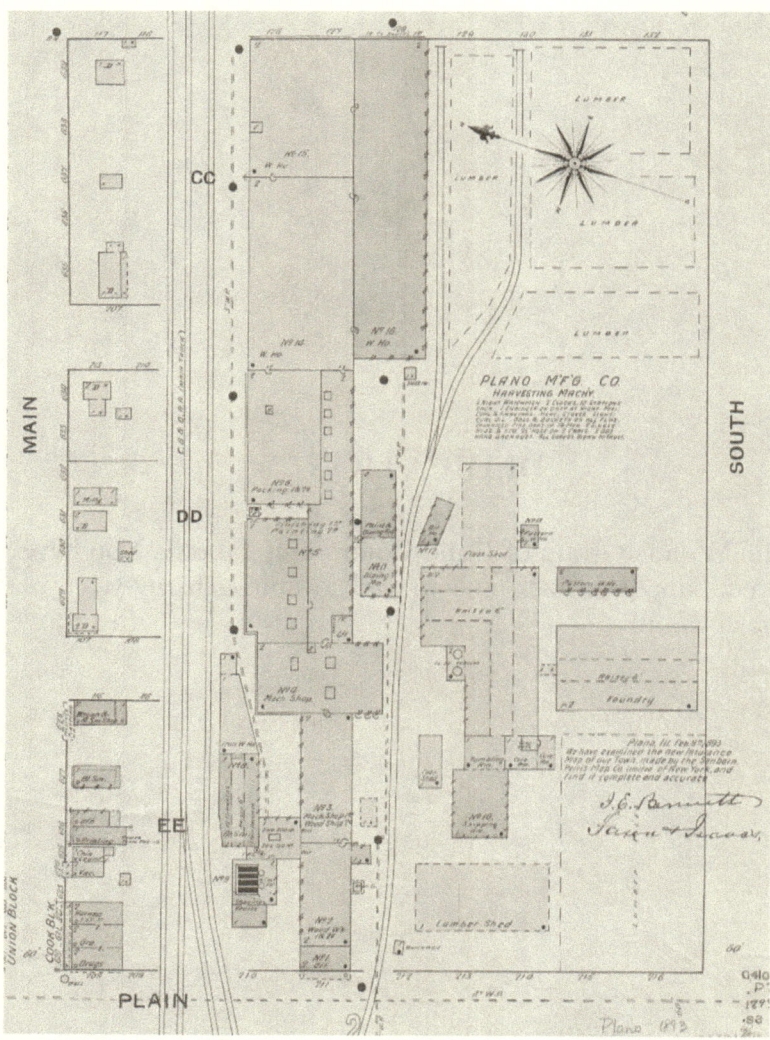

Plano – 1893

CONTENTS

CONTENTS vii

ACKNOWLEDGMENTS ix

PREFACE xi

CHAPTER 1 1

CHAPTER 2 4

CHAPTER 3 12

CHAPTER 4 29

CHAPTER 5 39

CHAPTER 6 49

CHAPTER 7 60

CHAPTER 8 79

BIBLIOGRAPHY 111

INDEX 113

ABOUT THE AUTHOR 115

ACKNOWLEDGMENTS

Thank you to Anne Sears and Kristy Lawrie Gravlin for assigning me the chapter on the harvester factories when we collaborated on the book, *Plano (Images of America)*. It sparked my interest in Plano's harvester manufacturing history. To the Plano Community Library Writers Group members I owe my thanks for their ever patient critiquing. Many thanks to Rebecca McNabb for suggesting the book's title. Most especially I want to thank my daughter, Carolyn Valentine Blakley, who created the cover for this book.

PREFACE

This small book springs from my fascination with a crucial part of Plano's early history: the invention and production of grain harvesters in the town where I've lived half my life. My research often led me down rabbit trails to interesting stories that did not relate directly to Plano. Mostly I resisted the urge to include them here, so you will not read about Charles Marsh's impressions of Abraham Lincoln and Stephen Douglas when he attended their debate in Ottawa, Illinois. Nor have I discussed the McCormick family feud about whether Cyrus McCormick or his father actually invented the reaper. And while two companies who began in Plano were among the firms that merged to form the International Harvester Company, I omitted the merger from my writing since neither company was located in Plano at the time of International Harvester's inception.

Plano is the main character in the pages that follow—the struggles of the people involved in the invention of the harvester, the factories and their history, and how these shaped the history of the town. I hope that this book will be engaging, whether the reader's primary interest is Plano itself or historical farm implements.

CHAPTER 1

Fred Ehrman shot himself through the heart at 9:00 on the evening of June 14, 1909. After he'd told his wife that she wouldn't see him alive again he took his revolver out to the shed behind his Seneca, Illinois home. Mrs. Ehrman hurried to a neighbor's house to ask for help but the sound of the shot rang out before they reached the shed. Described as a square, upright man who always paid his debts, Ehrman was to be interviewed by a United States postal inspector the following day. The U.S. Postal Service had begun an investigation into the Independent Harvester Company of Plano, Illinois for alleged mail fraud and Ehrman claimed to have lost $5,000 when he traded his Seneca hotel to the company for stock and the promise of a job.

Ehrman's suicide was a tragic consequence of his association with the manufacturer of farm implements that called itself a co-operative firm, its stock owned by thousands of farmers around the country. Four years after Ehrman's death everything finally came to a head.

Farmers Will Make War on the Harvester Company

The May 3, 1913 *Indianapolis Star* headline referred to the 27,000 farmers who owned stock in the Independent Harvester

1

Company and had sued for alleged mismanagement and extravagance. The federal government's investigation into the alleged mail fraud continued. For months farm magazines had discouraged investors from buying the company's stock and rejected their advertisements.

The designation "Birthplace of the Harvester" was a source of pride for the two thousand residents of Plano, Illinois, a farming community 50 miles west of Chicago. Fifty years had passed since their prized invention first emerged from the little Marsh Harvester factory, itself the earliest in a succession of farm implement manufacturers that provided employment for many Plano residents and equipment for farmers across the country and beyond.

Independent Harvester had used Plano's harvester manufacturing reputation to promote itself but now the town which had stood by the farm implement factory was in danger of being known as the home of the harvester stock fraud scheme. Successful reorganization of Independent Harvester was imperative.

By 1913 the Independent Harvester Company, established in 1905, had finally begun producing a moderate number of harvesters. But the management's imprudent spending, the sales agents' misleading promises, and the nonexistent stock dividends were too much for the stockholders. Their lawsuit and complaints to the federal government now put company president William Campbell Thompson and his officers on the hot seat. Independent Harvester's salesmen had promoted the company as organized to "buck the harvester trust off the map," in reference to International Harvester, the conglomerate of five manufacturers, that produced 85% of harvesters manufactured in the United States. But the Plano firm appeared to spend far more time and effort selling stock than on manufacturing farm equipment.

In hopes of satisfying their accusers, Independent Harvester's officers and board of directors resigned at a June 24, 1913 meeting in Plano. It was left to the newly elected officers, led by William Deering Steward, to redeem the company and the reputation of the town his family had helped build.

Steward, president of Plano State Bank and for fourteen years the town's mayor, gave up both jobs and took on the challenge of the Independent Harvester Company. Though his late father, Lewis Steward, had been intimately connected with the success of harvester manufacturing in Plano, "Deering", as the younger Steward was known, had no such experience. Turning around the troubled harvester company may have been the biggest challenge he'd ever face.

CHAPTER 2

*Go up the right hand creek and you will
come to the Garden of Eden*

In the 1830s, families from New England made their way to Kendall County, Illinois and settled in what is now Plano. The Steward and Hollister families were directed by a soldier who described the area to Marcus Steward and drew a sketch showing Chicago and the Fox River. He instructed Steward to follow the Fox River to a place where two small creeks emptied into the river, saying, "Go up the right hand creek and you will come to the Garden of Eden." The families followed the instructions, arriving in June 1838 after a six-week journey, and settled between the two creeks. Early descriptions of the locale are of fertile prairies with abundant timber, between clear silvery creeks.

Like others who came from New England, Marcus Steward and John Hollister discovered a crucial difference in farming the Illinois prairies. Preparing and planting the prairie required less time and effort than did the rocky New England fields. How easy it was to plant far more than a farmer and his available help could harvest during the short time in which the grueling work must be done. Finding a way to harvest their fields more efficiently became a priority. In Illinois, "more than elsewhere," wrote Steward's son, John, "one could sympathize with the efforts on the part of inventors to relieve the laboring classes from the stings of the stubble field and the scorching rays of the harvest sun." (*The Reaper, a history of the efforts of those who justly may be said to have made bread cheap,* John F. Steward)

The invention of the mechanical reaper revolutionized grain harvesting, though twenty years passed from the first U.S. reaper patents in the 1830s to their manufacture in significant numbers. With the reapers, harvesting with cradles and sickles was replaced by a horse-drawn machine to cut the grain and deposit it on the ground for farm workers to gather and bind by hand. The number of acres of grain that could be harvested in a day greatly increased but the back-breaking gathering and binding remained, motivating farmers to seek further improvements to ease the labor and reduce the number of workers needed. Numerous men in and around Kendall County and in other parts of the country experimented with improvements to reaping machines.

With his sons and Hollister, Marcus Steward spent many hours working on plans for a reaper. They constructed a model in a room at Hollister's Plano home but an actual reaper based on the model was never built. Instead, Steward rebuilt a reaper he had obtained in 1843 from Lott Presher of nearby Long Grove. Presher, a machinist, was known in the area as an inventor and manufacturer of reapers.

Steward added parts from a threshing machine and made other modifications to the reaper, including using hand saws for the cutting blades. Horses on each side pushed rather than pulled the machine. This reaping machine successfully harvested not only the Steward fields but those of the neighbors. Improvements were made to the machine over several winters. Steward's machine caught the attention of Cyrus H. McCormick who had patented his father's reaper and was building a reaper factory in Chicago. He visited the Steward farm in Plano around 1847 to spend several hours watching Steward's reaper in action. McCormick immediately threatened Marcus Steward with an infringement suit to which Steward replied that he probably understood the patent laws of the country as well as Mr. McCormick did. Steward's machine continued to cut grain for more than twenty years. He never patented it.

The day will come when men will not be so foolish as to throw their grain on the ground and then tear their hands in the stubble while getting it up again.

This comment by Thomas Judd of Sugar Grove, Illinois inspired Augustus Adams of nearby Sandwich to invent an improved reaper with a binder's platform. Adams's device, patented in 1852, worked well enough, but it never caught on and Adams turned his attention in other directions. A similar idea formed the basis of what became known as the Marsh Harvester and led to Plano's nickname, The Birthplace of the Harvester.

Charles Wesley Marsh and William Wallace Marsh emigrated from Canada to DeKalb County, Illinois with their parents in 1849. Charles was 15 years old and his brother, known as Wallace, was 13. During their childhood their father became a Second Adventist. He followed the preaching of William Miller who originally predicted the second coming of Christ would be in 1843. Unlike many of Miller's followers who fell away after the date came and passed, the senior Marsh continued to believe in Christ's imminent return. In consequence, it fell to his wife, and later also to his sons, to properly manage the family's finances and farm. Charles and Wallace began helping on the family farm at a very young age. When they first helped with harvesting, the scythe and cradle were their tools. Soon after the family arrived in Illinois, reapers began to appear in the area. In 1856 the Marshes bought a Mann reaper built by Haskell, Barker & Aldridge of Michigan City, Indiana.

On the hot, sunny afternoon of August 7, 1857, Wallace realized he could bind a sheaf of wheat faster than the reaper could cut and drop it on the ground. He and neighbor Simmons Brown were gathering and binding the wheat that day as Marsh's brother Charles ran the reaper and operated its revolving rake on the Marsh's farm in Shabbona Grove. Their Mann reaper had a canvas carrier or apron which brought the cut grain to a receptacle

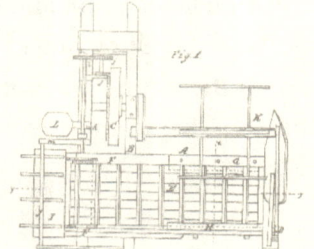

Top View of the Mann Reaper

from which it was dropped in gavels (the amount of grain that is bound into a bundle) onto the ground. Workers went from gavel to gavel, gathering the grain and binding it by hand into bundles they would then stand on end, several forming a shock.

At supper that August evening, 21-year-old Wallace told his family how quickly he could bind a gavel of grain. He was convinced that two people could ride on a modified reaper, alternately take the gavels of grain from the machine's receptacle, bind and toss them off, and still keep up with the pace at which the reaper delivered the grain to them. Wallace and Charles, along with their parents, Mr. Brown, and neighbor M.B. Richardson, discussed the idea throughout the meal. The prospect of saving time by binding right on the machine, and avoiding the constant stooping to gather the grain from the stubble excited them, though Charles contended that three people would be needed to bind. Brown supported Wallace's position. They decided to test the idea the following day.

Wallace had to help at another farm on August 8th. So, with Charles operating the reaper and delivering the grain onto the

7

ground with great care, keeping the gavels close together, Brown and Richardson ran from gavel to gavel, binding as quickly as they could. Four acres were bound in four hours, an excellent result that convinced them that two binders could ride on a machine and keep up with the grain as it was cut. Upon his return that evening, Wallace was elated to hear the day's results.

The Marshes were thinking of an attachment for the Mann reaper—a platform for the men to stand on and surfaces on which to rest the bundles while binding. They quickly realized that the Mann delivered the grain in the wrong position so they set aside the project until after the 1857 harvest was complete.

The Marsh homestead in Shabbona Grove

Over the winter of 1857—58 they worked on the plan for what they called a harvester. They devised the binding platform, the carrier to elevate the grain and place it onto the binding platform, and the placement of each element in order to balance the machine and distribute the weight properly, with the plan that two men would bind the grain while riding on the harvester.

In the spring of 1858, Charles went to Little Rock, Illinois where Samuel Bartlett helped him make a model of the machine. The Marsh brothers patented their harvester on August 17, 1858.

Charles had already begun construction on the first machine in June with the help of their uncle, Albert Hinds, farmer, wagon repairer, and sometime blacksmith, who lived nearby. Charles Marsh recalled, "We built the machine out of the Mann reaper and such castings as we could find. With it we cut our harvest in 1858, some 50 acres, and did good work, binding easily full cut after a little practice."

Wallace described that first harvester, "We thought on account of the elevator having to do the most work it should run faster than the lower canvas, and the machine was built accordingly. We started it up in Timothy [a perennial grass used mainly for hay] and the Timothy went up heads first and clogged the machine down. I saw the trouble at once and ran to the barn and got a lot of old belting and nailed it on the elevator pulley, which was of wood, increasing the motion of the platform canvas until it ran faster than the elevator. Then we cut the eight acre piece, you might say, without stopping."

...by showing him a goodly roll of money...

In early 1859 Wallace Marsh went to Chicago to begin their next harvester. He described the trip, "I must here give a little sketch of that Chicago trip for I must have been about as green as most other western country boys, with my high boots, home-made straw hat, and if I remember rightly, linen suit. I could not get any machine shop to build the machine for me, but they kept sending me from shop to shop until I landed at the foot of West Washington Street, between Canal Street and the river, in a low shanty-like building, with iron and wood-working tools, with power, about five o'clock, tired out. I finally got the owner interested by showing him a goodly roll of money, which made his eyes shine. I made a contract that evening for two machinists, one wood-worker, and the other an iron-worker, at $1.50 per day, each ten hours, and I was to pay him $1.50 per day for room, power and tools and he would help me make out specifications free... The boss helped to get a plan of what I wanted on paper... and then he advised me to go to Atkins & Wright's reaper shop and get guards, sections and what castings I could use, as I could get them at one-half price, which I did, getting back before night with a sack of iron on my back. I used that sack several times after that for the same purpose... That harvester we used on our farms that summer...we cut in all, 134 acres."

A foolish undertaking

The 1859 machine worked so well that neighbors ordered a dozen harvesters for the harvest of 1860. Charles Marsh would later describe their decision to build twelve harvesters that year as a foolish undertaking. Said Charles, "In the winter of 1860 we built a little shop and commenced on twelve machines... The castings were made, some of them in Chicago and some in Sandwich... In "assembling" the parts in our shop, where we did the work, we found that the machine work had been poorly done, and the machines came together badly. We patched, braced and bushed, but they were a poor lot. At the request of our customers we gave a public trial in a field of barley on the fourth of July... We had invited a number of 'eastern capitalists' to come and witness a trial of the machine. They came, they saw, and the thing broke down... these capitalists, skeptical and disgusted, left here to take the train back east."

Witnesses to the trial included a representative for Cyrus H. McCormick. His reaction to the new-fangled machine was that it resembled a "cross between a windmill and a threshing machine."

Charles and Wallace lost their confidence. They had mortgaged their farm and borrowed all the money they could to pursue the invention and now they were broke. They decided they could go no further. The neighbors refused to buy the machines and their friends told them to give up on the harvester. Said Wallace, "Sicker boys never lived than my brother and I after that trial."

Marsh Harvester of 1860

CHAPTER 3

If it can be made to run ten rods, it can be made to run ten miles, and there is a man in Plano that can do it.

Charles Marsh recalled, "Wallace and I spent some time in examining the machine and then went over to the rail fence along the road side under the shade of a tree. We sat down on the top rail of the fence. We didn't talk much but we did a lot of thinking, the result of which was that we were licked, we could borrow no more money, from our friends or the bank, the mortgage on our farm was almost due and we were so discouraged that we decided we were licked. Just at that time a man drew up his horse and buggy along the fence and alighted."

The man was Lewis Steward, oldest son of Marcus Steward and, at thirty-five, a successful farmer and businessman in Plano, Illinois. Steward was returning to Plano from a visit to tenants of his in Lee County. As he described it, "I guided my horse over toward the fence just back of the two young men and alighted from the buggy and went to them. At the near edge of the field a strange looking machine stood. I was curious. The two young men returned my greeting and upon question as to what sort of machine that was at the edge of the wheat, one of them replied, "Well we thought it was a harvester but we are about convinced we were mistaken. We don't know what it is." I then climbed over the fence and examined the machine.... Leading backward was a swath of cut grain. This swath was probably 25 or 30 rods in length. Alongside of this swath were bundles of grain bound with some of the wheat straw. I returned to the two young men and began questioning them as to why they didn't continue cutting and binding the grain."

After Charles Marsh explained about the failed trial, Steward concluded, "From what I see of your machine, the principle is sound but the construction is faulty. That machine has cut 40 rods successfully and as it has cut that much it can cut much more." I have an uncle in my home town at Plano who is one of the best mechanics in the world. He is a genius besides being an expert mechanic. I have an empty factory building and a power plant. If you will load your machine into a wagon and bring it down to Plano, I will pay all the expenses including the services of Mr. Hollister. You can board or rather live at my house. We should be able during the coming year before the next harvest to improve the machine and perhaps add some new ideas. After this, if the thing is a failure you are out your time and I am out my money. If it is successful we will form a partnership. I will take one-third as my share of the company. Later we can work out details."

The next day the Marsh brothers hauled their machine to Steward's factory in Plano. Wallace Marsh stayed in Plano that winter of 1860—61, working on the new machine with Steward's uncle, John F. Hollister. What would be the first successful Marsh Harvester was ready for trial by harvest time 1861. When the grain began to ripen, Lewis Steward's youngest brother, twenty-year-old John Fletcher Steward, took the harvester out to the family homestead and commenced the trial in a field of rye. That first machine cut 164 acres in 1861 and was used for over a decade.

His blood is up instanter!

Charles Marsh tried unsuccessfully to interest others in the venture. Even an impressive showing at a field trial at Clark Barber's farm, two miles north of DeKalb in 1863 brought no interest. The field trial was held on July 15 and 16 as reported by the *Chicago Tribune.*

"The Marsh Brother's hand binder. This is a novelty, and attracts today, as yesterday, a great deal of attention. My description must be brief. The machine is drawn by two horses, driven by a driver who sits elevated high over a large driving wheel, the gearings of which propel the sickle, reel, and an endless apron on which the grain falls as it is cut, and is carried over an upper cylinder, at the right of the driving wheel, and deposited in a trough. On a platform beside this trough stand two men, who bind the grain as fast as it falls there—each alternating with the other in binding a bundle. The machine has been once around, both men binding. Now one of them quietly sits down on his end of the platform, folds his hands, and the other proceeds to bind all the grain cut, as fast as it is delivered to him from the platform of the reaper. It is true, the width of swath is not great, neither is the team driven rapidly; but the work is well done. One man does it. It is pretty snug work, and somebody suggests that he cannot do it all day. His blood is up instanter! He offers to bet that he can cut, and bind, alone, with a man to drive for him, twelve acres in twelve successive hours. Nobody takes the bet; but sundry neighbors who know what he can do, cry, 'Yes, sir, and he'll do it, too. How long was he cutting and binding this acre? Fifty minutes.' There is data upon which to bet. There was no "hurrying to and fro" in the performance of this work. No better stubble—no cleaner, and snugger bundle can be found on the field cut today."

The Marsh brothers won "Best harvester and hand-binder" for which they received a premium of $5.00 and a diploma but brought no investors nor any manufacturers eager to take on production of the new invention. Despite their discouragement, the brothers were glad that their neighbor was eager to borrow the harvester in the evenings to harvest his grain and that their father stopped suggesting that the harvester could be used to block up holes in their broken fences.

Meanwhile, the American Civil War raged on. Young men headed off to fight leaving fewer on the farms and accelerating the need for more efficient ways to harvest grain. Farmers bought reapers and although the Marsh brothers' harvester was still in its infancy, the need for their labor-saving device had intensified.

When the Civil War broke out, Lewis Steward's brother, George, happened to be in the South. George, a skilled mechanic, was in New Orleans overseeing building construction in April 1861 when Fort Sumter was fired on, having been sent there by his employer, a Holly Springs, Mississippi foundry. He served briefly in the Union Army under General Sherman before returning from New Orleans to Holly Springs. Working at the foundry once again, making cannons for the Confederacy, he decided it would be prudent to conceal his earlier service in the Union Army. Anxious to return to Illinois, Steward made his way north, first in a Union army wagon and then in a box car, finally reaching home in July 1862.

Before his sojourn in the South, George Steward had operated a small plow and wagon factory. Upon his return, Lewis Steward proposed production of the Marsh Harvester in Plano since Charles Marsh had not found any alternatives. With Lewis Steward financing the operation, Charles Marsh and George Steward commenced production of 50 harvesters in Plano in a small stone shop connected to the grain elevator owned by Lewis Steward and Gilbert Denslow Henning. They called the company Steward and Marsh. The shop had previously been used as a sorghum mill and then as a sash and door factory by the firm of Latham and Doty.

The little company began work on 50 harvesters, 26 of which they completed for the 1864 harvest. Wallace Marsh took one of the harvesters to a field trial in DeKalb. Reaper and mower salesman John D. Easter was impressed with the performance of the new invention. When he related what he had seen to his partner Elijah H. Gammon, they decided that their Chicago firm, Easter and Gammon, should take a chance on the harvester.

Meanwhile, although all 26 of the Marsh's harvesters were sold and worked well, they had cost almost as much to produce as the sale price. The cost of machinery and materials scared the Marsh brothers. So when Easter and Gammon visited the Marsh farm in Shabbona Grove later in 1864, the inventors were ready to grant them a license to manufacture and sell the Marsh Harvester for a $7.00 royalty per machine. The brothers also accepted an offer to sell a one-third interest in their patent for $5,000 to Champlin & Taylor of Sycamore, Illinois.

Easter and Gammon's license permitted them to sell the harvester in Illinois, Indiana, Michigan, Iowa, Wisconsin, and Minnesota. They contracted Parker & Stone, of Beloit, Wisconsin, to build 100 Marsh harvesters for the harvest of 1865 but only a small number of the machines were well enough constructed to be put on the market.

The Marsh brothers still had to run their farm in DeKalb County so in the fall of 1864 Charles returned there from Plano and Wallace went to Plano to work at the fledgling factory. As Lewis and John F. Steward both noted, of the pair, Wallace was more the mechanic while Charles was better qualified to take care of their business affairs.

Over the next three years, 1866–68, two additional companies were licensed to manufacture the harvesters, an arrangement the Marshes would regret. The quality of the machines produced by these companies was uneven at best, forcing the Marshes to

continue their participation in the Plano factory in order to pre-
serve the good reputation of their harvester.

When farmers bought a harvester, it had to be "set up" for
them initially and some instruction given so that they could learn
how to start and successfully use the machine. Charles Marsh
wrote of an early experience in Iowa. He had gone there in 1865
to set up and start three new harvesters. After the preparation out
in the field, Marsh demonstrated the harvester, performing the
binding for those gathered to watch. A large well-muscled man in
the crowd demanded to give it a try. According to Marsh, the
fellow was intoxicated, making the task of binding while riding on
the harvester more challenging. He gave up and was soon thoroughly embarrassed when a young woman asked to try binding
and did it quite well after a little instruction from Marsh. (*Recollections 1837–1910*, Charles W. Marsh)

Young John F. Steward, the brother who had launched the
first Plano-built harvester in 1861, joined the Union Army soon
thereafter and did not return to Plano until 1865. He then
jumped right into work at the harvester factory, discovering an
operation where funds were tight and machinery rudimentary.
He recounts the early challenges of building harvesters in the
makeshift Plano factory:

> *"Practically all work was done by hand. Money was
> scarce; ear corn was at times cheaper than coal. Even as
> late as the winter of 1865–6 I shoveled thousands of
> bushels of the best Yellow Dent corn into the furnace
> to keep the wheels of the improvised factory turning.
> Every bolt and nut for the machines except the most
> common sizes, was forged and threaded by hand. All
> holes in wrought metal parts were drilled in a lathe or
> upon a wooden drill press. We had no punch press of
> any kind. Two circular saws, a rip and crosscut, worked
> out, in the rough, the wooden parts. A single planer, and
> that of an early type, aided materially, but every part of
> the machine had to be finished by hand labor at the*

bench. The younger Mr. Marsh was particularly active. Wherever he went, and wherever the machines had a chance to reveal their powers, prejudice was overcome and public incredulity slowly faded." (The Reaper, a history of the efforts of those who justly may be said to have made bread cheap, John F. Steward)

The crude method of construction meant that there was little profit from the $225 to $300 the machines sold for. Much time was expended in sending John F. Steward, Wallace Marsh, and others here and there repairing the "Old Blues", as those first harvesters, painted blue, were called. The difficulty, noted Robert Ardrey in *The Iron Age*, was that "a machine which works well in the hands of its inventors, who know how to handle it and keep it together, is a different proposition from the construction of machines that are to be put out in the hands of farmers, who are seldom skilled mechanics, and who will drive a 'new-fangled' invention into the fence corner and have nothing more to do with it if it gives them any trouble."

Both the Plano company and the firm of Easter and Gammon spread the word about the harvester. An April 16, 1864 *Prairie Farmer* article, reprinted in the *Kendall County Record*, explained how the machine worked and detailed variations in its operation:

"But further we have arranged so that our harvester can be raised to a high cut; the concave dropped out of the way, and an apron added which will take the grain from where it is delivered to the binders and deliver it into a wagon, like an ordinary 'header'.

"We think the advantages are obvious. There is great economy both in labor and expense to the farmer in binding, and when the operators get tired of binding, or when the straw become brittle, by simply raising the machine and putting up an apron, we have a perfect head-

cutter, cutting 5 ft. 4 inches wide, easily drawn by a span of horses."

In the summer of 1865 Lewis Steward, always with his hand in a variety of enterprises, and with an eye to aid the harvester plant, built an iron foundry in connection with one of his businesses, Steward and Henning. The 120-foot-long stone building, noted the *Kendall County Record,* had a cupola (furnace) and "all the paraphernalia of a foundry and machine shop, where farmers and others can have their repairing done at short notice. Arrangements are being made for the manufacture of the celebrated Marsh reaper this fall. The foundry will probably be in operation in the fall."

While the construction of the foundry continued through the summer, the firm of Marsh and Steward found ways to use their factory to bring in additional revenue. They ran an advertisement in the *Kendall County Record,* beginning that April offering to repair "all kinds of Reapers and Mowers in the best possible manner at short notice and warranted satisfaction or no pay. In fact...all kinds of agricultural implements such as reapers and mowers, horse-powers and threshing machines. Grain drills, corn planters, cultivators, plows, carriages, harrows, wagons, etc." The ad, signed by Lewis's brother, George, also offered, "Blacksmithing in all the branches promptly and neatly done. Tire setting done without cutting and welding, thereby producing a better job at a cheaper rate. We have now in store and shall keep constantly on hand, for sale, an assortment of wagons, which are not surpassed either in quality or style or price by any in the market, which will be fully guaranteed, as well as all work turned out of our shop."

...mutilating his face in a shocking manner...

On August 14th, 1865, John F. Hollister, the Plano mechanic who had helped the Marsh brothers make their invention a success, had a dreadful accident. While working in Steward's grist mill, he was caught by a belt and thrown nine or ten feet, striking his head and face. The local newspaper described the accident, "mutilating his face in a shocking manner, also fracturing both bones in the left arm near the hand, also the radius in the right arm. It is with great difficulty that he can swallow or move his head in any way, apparently from concussion of vertebra and the chord (sic)." Despite the reported severity of his injuries, Hollister recovered sufficiently to return to work and lived another seventeen years.

Steward's Grist Mill on Main Street, Plano

John Fletcher Hollister
1811-1882

John F. Hollister and family migrated from Pennsylvania to Kendall County, Illinois in 1838 with the family of his brother-in-law Marcus Aurelius Steward. Hollister was an expert mechanic and a poet; he published a book of poetry, *Sunflower*, in 1881. Hollister gave Plano its name which derives from the Latin, due to its location on a plain between Little Rock and Big Rock Creek.

After Hollister assisted the Marsh brothers with the invention of their harvester in 1860, he worked in the harvester factories in Plano for many years, devising a number of improvements.

John Hollister was, "...a strong advocate of temperance, unalterably opposed to the subjection of women, the uncompromising foe of all forms of sensuality, and of the follies of dress and manners. He had a great aversion to humbugs and pretenders. He was positive in his convictions, fearless in expressing them, and scrupulous in carrying them into effect." *(The Hollister Family of America; Lieutenant John Hollister of Wethersfield, Connecticut and his descendants*, compiled by Lafayette Wallace Case, M.D. 1886)

John Hollister's two children died in infancy. His wife's nephew, Albert H. Sears, spent much of his childhood living with them as their foster son. John Fletcher Hollister died February 11, 1882, at age 71.

The challenges of building and selling the harvester were great but well worth the effort. The demand for the labor-saving machine continued to grow. One hundred harvesters were manufactured by Steward and Marsh for the harvest of 1866. With farmers taking notice of the new invention, in May 1866 a reporter for the *Kendall County Record* toured the factory:

> *"Mr. George Steward very kindly left his work and showed us through the factory. Everything is in complete order and the work done with perfect system. The machinery is driven by a large steam engine and shafts and pulleys run all through the building. On the ground floor is a planing machine for the dressing of plain boards, such as flooring; this is in almost constant requisition. Upstairs are the drills and lathes for the iron work, and saws, lathes, and planers for the wood work. One planer is very peculiar. It is constructed that it will dress lumber in any required form, and is one of the greatest labor-saving machines ever invented. Its working is truly marvelous. The paint shop is in the east end of the building.—After the harvest season is over, another story will be added to the building, a cupola built, and a foundry be carried on to make their own castings. Between 25 and 30 hands are employed in this factory. The warehouse of Messrs. Steward & Henning adjoining, is very capacious, and all the work is done by steam power driven by the same engine that is in the shop."*

...amputation was necessary

That same week the *Kendall County Record* reported "A serious accident occurred at the machine shop of Steward & Marsh in Plano, on Friday last, resulting in the loss by Mr. Marsh of the two first fingers from the left hand. Mr. Marsh was using a machine called an irregular planer, and upon placing a board which he held in his hands against the machine, the board was drawn rapidly towards the machine bringing his fingers in contact with the knives, mutilating them in such a manner that amputation was necessary." While the report didn't specify which Mr. Marsh was injured, it is likely that Wallace Marsh was the one present at the factory that day.

By the summer, to build the 100 harvesters for the 1866 harvest season, Steward and Marsh employed 50 men. They announced that an additional story would be added to the factory in the fall so that they could manufacture up to 500 harvesters for 1867. An improvement to the 1866 model harvester was the ability to cut grains of various heights. Whereas the first Marsh harvesters were not adjustable, the 1866 model could be adjusted "by lifting the machine relative to the main supporting wheel and there securing it." (*The Reaper, a history of the efforts of those who justly may be said to have made bread cheap,* John F. Steward) Before the year was over, Wallace Marsh had officially joined the firm which was then renamed Marsh Brothers & Steward.

For 1867, one hundred fifty harvesters were produced at the Plano factory. John D. Easter (no longer partnered with Gammon) under a license from the Marshes, had 400 Marsh Harvesters manufactured by Warder & Mitchell of Springfield, Ohio that year as well. The Warder & Mitchell machines turned out badly due to a fire at the factory and poor workmanship. Warder & Mitchell also built harvesters for the firm of Gammon & Prindle that year and the next.

A significant source of employment in the Plano area, Marsh Brothers & Steward was frequently mentioned in the *Kendall County Record.* As 1867 came to a close, the newspaper reported that the "factory at Plano has a new steam whistle to call the hands from labor to grub, and from grub to labor. The whistle has rather a weak voice but it's very young."

Marsh Brothers & Steward began 1868 with plans to build 450 machines, to be sold for $225 each, but they only completed 250. As in 1867, Wallace Marsh was superintendent of the factory.

George H. Steward directed the mechanical work. "They have now in operation a large iron drill that was made in the shop—cast and finished entire—which would be a pride to the best machine shop in the country. They are now making a machine punch for punching large plates of iron in which is one casting weighing 900 lbs., cast at the Plano foundry." (*Kendall County Record*, March 19, 1868)

In May 1868 the company advertised their prices in the *Kendall County Record*:

> *"As there have been some misrepresentations and misunderstanding in reference to the price of the "Marsh Harvester", we hereby announce that our price and terms are as follows: $220 half cash, half in 6 months, at 10%. $210 all cash. At these prices we give full warranty and hold ourselves in readiness, to see that the machine is properly put up and started at our expense. To a person who we know as capable of managing the machine, taking it without warranty, or "man to put up and start", paying cash, we will discount $10 on cash price. The same prices and terms, freight only added, will be observed at each agency. Marsh Bros. & Steward. Plano. May 19, 1868."*

By the end of the year the Marsh brothers decided that they could finally withdraw from personal participation in the firm and return to DeKalb County. The name was changed to Marsh, Steward, & Company.

The Marsh Harvester had caught the attention of the McCormick Harvesting Machine Company in Chicago. At first considered a "humbug" by the McCormick company, the harvester was now taken seriously, though it took McCormick five years to get into the market. Also beginning about this time were the many lawsuits over patents and licenses involving Marsh, McCormick and a number of other companies.

In June 1869 sixty men labored in the Marsh, Steward & Company factory. They had completed 750 harvesters for the season and sold most to Elijah H. Gammon who was still selling farm equipment, but without his partners of previous years. The Plano factory was now 150 x 50 feet with an elevator of 60 x 50 feet and a 40 horsepower engine to power their works. They began building a 60 x 50 foot warehouse that summer. "...the business done this year amounts to $150,000 and the capital invested is $100,000 which will be greatly increased the coming season. (*Kendall County Record,* June 24, 1869)

William Wallace Marsh

Charles Wesley Marsh

Charles Wesley Marsh   William Wallace Marsh
1834-1918   1836-1918

Charles Wesley Marsh and William Wallace Marsh, natives of Canada, settled with their parents and sister in DeKalb County, Illinois in 1849. The family purchased a farm in Shabbona Grove where the brothers invented their harvester.

After selling their interest in the harvester factory in Plano, Charles and Wallace Marsh established the Marsh Harvester Company in Sycamore in 1869. The brothers sold the successful company to John D. Easter in 1877. When Easter's company failed very shortly thereafter, the Marshes, honorable men with concern for the well-being of others involved, agreed to take over. They assumed the debts of the company and resumed building harvesters. With automatic binders entering the market about this time, the plain harvesters were losing ground. On top of that, their original patent turned out to be defective due to bad advice regarding its wording. They closed Marsh Harvester Company and opened Marsh Binder Manufacturing Company in 1881 but with the weight of the old debts and patent issues, the company failed in 1884. Charles and Wallace Marsh lost the fortunes they'd earned.

The following year, Charles Marsh became the editor-in-chief of the newly established *Farm Implement News* which became a leading trade publication. He was elected to the Illinois legislature in 1868 and to the state senate in 1870.

In 1877 Wallace went to Nebraska and then on to Arkansas where he ran a lumber business and made good investments. He returned to DeKalb County in 1895.

At the urging of friends and family, Charles Marsh wrote and published his autobiography, *Recollections, 1837 – 1910*. The brothers died just months apart in 1918.

CHAPTER 4

The factory was humming in November 1869. Marsh, Steward & Company hoped to build up to 1,500 machines for 1870 but a need for sufficient capital was a constant concern. Help came in the person of Elijah H. Gammon who visited Plano that month and toured the harvester factory with an eye to investing in the firm that built the Marsh Harvesters he sold.

By January 1870, plans for the company's future were firming up. Gammon came on board and, with Lewis and George Steward, would run the factory. The *Kendall County Record* reported "authorized capital of $80,000." The harvester factory was located on the east side of Plain Street (later renamed Center Avenue, now commonly called Center Street), just south of the railroad tracks. An elevator, mill, and the harvester manufactury shared one large building. The youngest of the Steward brothers, John F. Steward, took over as superintendent of the factory.

Construction of 1,000 harvesters began for the 1870 harvest. The firm shipped three of the machines to Europe—one to the Austrian Minister of Agriculture, one to the Agricultural School at Ungarisch Altenburg in Hungary, and one for the International Reaper Trial to be held that summer at Grosswardein, Hungary. In May, Charles Marsh left for Europe, where he stayed through the summer, introducing the harvesters and participating in the reaper trial in July. His autobiography, *Recollections, 1837–1910*, provides a fascinating account of his trip, including the harvester's triumph at the reaper trial. The harvester's success resulted in shipments of fifty of the machines to Europe the following year, with another Steward brother, Amasa, and two additional sales agents traveling there on behalf of Marsh, Steward, & Company.

Sometime in 1870 an acquaintance of Gammon's from Maine came to Illinois. William Deering, a successful woolen merchant, sought to increase his wealth through investments in what was then "the west." The result of Gammon and Deering's meeting in Chicago was providential for both men. Gammon convinced Deering to loan him $40,000, providing much-needed funding for the factory in Plano. For Deering, it was the beginning of his second fortune. The $40,000 investment led to an $80,000 profit for the firm within two years and motivated Deering to become a full partner in the enterprise in 1872.

After early doubts about its reliability and worth, the harvester had finally gained acceptance. The demise of the reaper had begun. Others now sought to enter the harvester market:

> "In 1871 Mr. Ellward (sic) [refers to John H. Elward] had six machines built at Plano—a modification of the Marsh; and about this time W.R. Low got out his machine; which was a change in another direction. Ellward's machine became well known in time as the St. Paul harvester; and Low's, improved by himself and Augustus Adams, as the Low, Adams & French harvester, built by the Sandwich Manufacturing Company. Both these concerns recognized the rights of the inventors, took out licenses and built good machines." (American Agricultural Implements, Robert L. Ardrey, 1894, p. 60)

Marsh, Steward & Company built 1,400 harvesters in 1871 and expanded its facilities, replacing a shed east of the factory with a two-story stone building, 50 feet x 100 feet, to be used as a paint shop and store house. They ushered in 1872 with the addition of a 4,000 pound lathe.

Expansion of overseas sales continued with the shipment of two train cars of harvesters, destination: Russia. By spring the firm was renamed Lewis Steward & Company. The Marsh brothers were now running their own factory in Sycamore: Marsh Harvester Manufacturing Company.

John F. Steward took a sabbatical of sorts from his work as factory superintendent in 1871. Accompanying the second Colorado River expedition of Major John Wesley Powell, Steward served as assistant geologist from the spring through early November when ill health forced his return to Illinois. Powell and Steward had met as servicemen at Vicksburg during the Civil War, encountering one another while collecting fossils. Steward was wounded soon thereafter and it was the effects of this old injury that cut short his participation in the Colorado River expedition. In January 1872 Steward finally arrived back in Plano. He did not return to his position as superintendent of the harvester factory, instead moving to Chicago where he served as foreman of the adjustment department of the Wheeler and Wilson Sewing Machine Company. A Wheeler and Wilson sewing machine, propelled by steam power, was used in the Plano harvester factory to make the canvasses for the harvesters. In November 1872, John F. Steward left the company and his brother, Amasa, who had been the general agent for the sewing machine company in Iowa, became the superintendent of the adjustment department.

John F. Steward on the Colorado River Expedition
Glen Canyon 1871

Expansion and improvements at Lewis Steward & Company filled the second half of 1872. The harvester factory dismantled the west section of their warehouse and put up a three story brick building. The new building's basement was for storage and repairs; the first story housed the offices of both Lewis Steward & Company and those of the firm, Steward & Henning. The second floor contained a pattern room and sewing machines. The *Kendall County Record* noted that this was Plano's first three-story building. The company purchased a new engine and a new boiler to provide sufficient power to manufacture the 2,200 harvesters planned for 1873.

As 1872 came to a close, the *Kendall County Record* reported that the companies building Marsh Harvesters expected to manufacture a total of 5,000 for the 1873 harvest. This included those produced in Plano along with 1,350 to be built at the Marsh factory in Sycamore.

Lewis Steward & Company expanded its territory in 1873. George H. Steward traveled to the east coast where he combined his sales trip to seven states with a pleasure trip.

Marsh Harvester

Improvements to the factory continued with the purchase of a new drill press and the installation of an R.F. Sturtevant "noiseless blower" in the blacksmith shop in January 1873. That same month, Lewis Steward traded a harvester for a bell weighing more than a half ton for the belfry of Plano's Congregational Church, located on Steward Street between Hugh and West Streets.

By spring the firm had increased their expected output of harvesters. They also added two new iron turning lathes, an iron planer and a double surfacer planer. One hundred eighty men worked in the factory. When the 1873 machines were finished in June, the total was 2,900 from the Plano factory. Wood Department foreman E.C Fields and his men put the finishing touches on the final 300 harvesters the last week of June. A few weeks later, the *Kendall County Record* reported that Fields "was last Saturday presented by the men in his department and under his supervision with Webster's Pictorial Unabridged Dictionary, a copy of Shakespeare's complete works, bound in Morocco, and a large fine Album, the whole costing over $30."

Work for the 1874 season was underway by summer's end with plans to build 4,500 harvesters. The *Kendall County Record* noted that "our town is rapidly filling up with laborers and mechanics, making houses scarce and rents high. A few of our citizens are endeavoring to meet the emergency by putting up some very nice and commodious dwellings." September saw the construction of a foundry, paint shop and finishing rooms, almost doubling the factory's capacity. Two hundred forty men worked at the factory in the fall of 1873.

Elijah Gammon's health had deteriorated in the summer of 1873. He contacted William Deering and asked him to come to Illinois for a few months to handle Gammon's role in the harvester firm. Deering had no experience with agricultural implements, saying later, "...I didn't know one of our own machines when I saw it. And I was embarrassed beyond measure that time I went first to buy pig iron. I said I wanted so many tons of pig iron. 'What kind of pig iron?' the dealer asked—and I didn't know!" But Deering came to Illinois and stayed, initially in Gammon's home in Chicago while Gammon and his wife traveled to Europe.

While Gammon was in Europe to improve his health, John F. Steward went to Europe as well in 1873, his purpose to promote the harvester. Among his stops were Vienna, London, and Odessa. Steward's efforts bore fruit. Russia, Denmark, England, and Scotland were now destinations for the harvesters.

The company revised its name as the status of partners changed, progressing from Lewis Steward & Company, to Gammon & Steward, to Gammon, Deering & Steward, and then Gammon & Deering. Gammon and Deering maintained their sales office at 193 Washington Street in Chicago. In January 1874 a local newspaper interviewed Elijah Gammon and William Deering there. They reported that the firm was selling the Marsh Harvester in several states, including Iowa, southern Minnesota, Michigan, Kansas, Missouri, Texas, and Ohio. The two businessmen described the popularity of the harvester and remarked that the demand was such that it was hard for the factories in Plano, Sycamore, and Rockford to keep up.

A global depression began in 1873, lasting until 1879. It is likely that the resultant falling wages led to the harvester firm's explanation in a *Kendall County Record* article that fall, "Mr. Steward has always said that the wages of laborers should be no less than the railroads were paying—which is $1.25 per day, a day constitutes 10 hours' work, but they are only running 9 hours at present, consequently that class of men only receive 9 shillings. Manufacturies are cutting down all over the U.S. Mr. Steward still pays his men as of old, and does not propose to cut down on wages."

Steward's name appeared often in the local news. His importance to Plano could hardly be exaggerated. The January 1, 1874 *Kendall County Record* averred that, "...like the town pump, he [Lewis Steward] is a fixture. He is cumbered with the care of 4,000 acres of land; has a railway on his hands, is President of the Chicago, Millington, and Western Road, is joint proprietor with two other parties of five of the best horses in Kendall

County, and has an interest in a tannery, a saw mill, a cheese factory, and perhaps a grist mill!"

By the end of June 1874, four thousand one hundred harvesters had been shipped for the season. Three hundred thirty men had labored at the factory to get out the largest number of machines the firm had ever produced. George, John, and Amasa Steward all traveled to various states in the interest of the harvester firm. In July, with the shipments complete, plans for 1875 began. The paint shop would be enlarged to accommodate the 6,000 harvesters to be built which exceeded the McCormick Harvesting Machine Company's output of 5,000 that year, it being the first year McCormick manufactured harvesters.

Elijah H. Gammon

Elijah H. Gammon
1819-1891

Elijah H. Gammon, a native of Maine, became a Methodist minister at age 24. Suffering from a bronchial ailment, in 1851 he moved his family to Illinois, hoping the climate would be favorable for his health. He served as a minister in northern Illinois, settling his family in Batavia. After only a few years, his health deteriorated such that Gammon was forced to give up his vocation, his throat difficulty preventing him from preaching.

Elijah Gammon entered the agricultural implement business in 1861, partnering first with J.D. Easter, and later with James Prindle, to sell harvesters. Gammon then joined the Plano firm of Marsh, Steward & Company which became Gammon & Deering. Gammon sold his interest in the company to William Deering in 1879, a difference in opinion regarding the direction of the company having resulted in ill-will between the two men. Elijah Gammon was involved with the Plano Manufacturing Company from the time of its inception, as a major stockholder and serving as Vice President.

Gammon founded Gammon Theological Seminary in Atlanta, Georgia in 1883, a Methodist seminary expressly for the education of African Americans. He enthusiastically supported the institution until his death in 1891.

CHAPTER 5

Various inventors had been working to devise an automatic mechanical binder since at least 1850. Not until the appearance of the Marsh harvester were successful automatic binders invented—as attachments to the Marsh machine—using wire to bind the sheaves. William Deering, eager to encourage inventors, provided Marsh harvesters to them at cost. Early on, twine had been considered for binding but a mechanism that would automatically tie a knot with twine had not yet been devised. Wire binding attachments could wrap wire around a sheaf of grain, cut the wire and twist it to hold the bundle of grain together. The wire was run from a reel, the driver regulating the size of the bundle.

The Withington and Sylvanus Locke wire binders, both attached to Marsh harvesters around this time, were successful. The McCormick Harvesting Machine Company had chosen to build the Withington binder, while Walter A. Wood Mowing and Reaping Machine Company manufactured the Locke binder. The Plano firm selected a wire binder invented by John H. Gordon in consultation with his brother, James. Since 1872, each year the Gordon brothers had brought to Plano the binders they were working on. Gammon, Deering, & Steward began building the Gordon binder in 1874. Despite Gammon's reluctance, Deering, who owned the largest interest in the business, had insisted that the company would build wire binding attachments for the 1875 harvest. He recognized that automatic binders were the logical next step in harvesting machinery. Improvements had been made to the binder; it had been tried out on a farm in Lee County, Illinois and used through that harvest season and Deering was anxious to demonstrate that the firm was willing to take risks in order to further innovation. Though it cost more to bind with wire than by hand, farmers wanted automatic binders so they could harvest their grain themselves.

The Plano firm's new binder was a disappointment at first. All 113 binders were returned to the factory. John F. Steward rebuilt them, replacing the holding and twisting devices and making

Gordon Wire Binder attachment

other improvements. They were then sold for the 1876 season along with 300 more wire binding attachments.

William Deering spent his 50th birthday, April 25, 1876, in Texas where the first of the rebuilt binders was being tested. Deering considered this point to be the beginning of his great success.

Marsh Harvester as sold by Gammon, Deering & Steward, 1875

In September 1875, Lewis Steward sold his interest in the harvester firm which then changed its name to Gammon & Deering. In February of the following year, Steward was nominated for governor by the Democratic Party as well as the Greenback Party which represented the interests of farmers. Steward resigned as president of the Chicago, Millington & Western Railroad, a position some thought inconsistent with support for farmers. Though Steward was an uncooperative, intractable candidate he lost to Shelby M. Cullom by only 6,834 votes, one of the smallest losses in an Illinois governor race in the state's history (by number of votes).

Upon Lewis Steward's departure from the firm, Elijah H. Gammon moved to Plano to help manage the factory. Steward sold Gammon over five acres of property on North Street on which to build a home. Before the end of September, work began on the Gammon residence. The home at 603 E. North Street was complete by spring of 1876.

For the 1876 season, Gammon and Deering advertised their Marsh Harvester, along with the new Chicago Harvester King, and an automatic wire binder. The Chicago Harvester King cut six feet wide as opposed to the Marsh's five foot cut. It was similar to the Marsh Harvester, but with binding tables adapted for two or three workers to bind the grain and several other improvements. Gammon and Deering informed their customers that the limited number of wire binders produced for 1876 would be followed in 1877 with as many as farmers desired.

The wire self-binder attachments to the Marsh Harvester gained popularity. Farmers could harvest 15 or 16 acres per day with the self-binder; the sheaves were bound securely and, at 20 to 30 cents per acre, the cost of wire was considered reasonable.

Factory employee Herman N. Kennedy, upon his retirement, reminisced, "When the wire binder was ready for manufacturing, the place began to boom. We made two punch and shearing machines and a large drill press, which are still in use at the Deering works. We also bought some new lathes and drill presses, and about 1875 the factory began to rush so that they had to start a night shift, which was done by advertising in Chicago for men, and getting what new machines were required. About this time they started to build the Warrior mower, and also to experiment on other styles of mowers."

In the spring of 1876, the firm provided employment to 216 men. Plano's population had reached 1,800. With the Plano factory bustling, Elijah Gammon, William Deering, and Lewis Steward met with the railroad superintendent to discuss increased side tracks and space for lumber yards. Soon thereafter the firm purchased the railroad's Plano warehouses and then traded them to the company of Steward & Henning in exchange for the latter's warehouses in downtown Plano. Gammon & Deering then replaced the frame warehouses with brick buildings, and, in conjunction with the construction, West Street was opened between John and Main Streets.

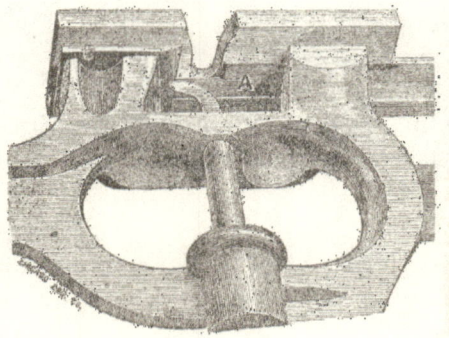

Twisting, holding, cutting device of Deering's Gordon Binder, 1876

Map showing parts of John and James Streets
taken over by the Harvester factory

Chicago, Burlington, and Quincy Railroad put in the new side track extending 900 feet along John Street east to James Street. James Street, at that time, ran continuously from at least six blocks north of the railroad to as many blocks south of the tracks. C.B. & Q.'s revenue from shipments through the Plano train station was $40,000 annually, with $24,000 of that related to the harvester factory. Gammon & Deering had purchased from J.D. Easter the rights to sell harvesters in Indiana, Wisconsin, and Illinois, brightening the future for Plano and its harvester factory.

In further preparation for construction of harvesters and mowers for the 1877 season, Gammon & Deering purchased an 85 horsepower engine from the Putnam Manufacturing Company of Fitchburg, Massachusetts. Installation of the $3,000 engine with its twelve-foot drive wheel attracted the attention of a sizable group of factory employees that October.

In the spring of 1877 the company was finishing construction of 1,200 Gammon & Deering mowers, the successor to their Warrior mower. The mowers, in addition to a large number of harvesters and wire binder attachments, were being readied for shipment to Texas, Iowa, Oregon, Utah, Missouri, Minnesota, and Canada in March.

After the summer slowdown, workers were called back, employment increasing from about 50 men in late August to 200 by the end of October. Construction of 2,000 Gordon Crane wire binder attachments, 1,500 mowers, and 3,000 harvesters was well underway by November, with large quantities of lumber and cast iron arriving daily. Orders for their products exceeded Gammon and Deering's expectations for 1878.

The company had begun manufacturing mowers in 1876. By 1878 their mower, its name changed from the Warrior to the G. & D. (Gammon & Deering) Improved Mower was advertised as combining the "most desirable points of a first class light draft grass cutter" that would "not clog in any kind or quality of grass and can be started when standing in stout or matted grass without backing up." (*Kendall County Record,* June 13, 1878)

The newspaper continued with a summary of Plano's harvester manufacturing history to that point, including the following:

"The management of our shops seems to be in competent hands; although this firm have large interests in Chicago, yet Mr. Gammon resides in Plano and gives much of his time and personal attention to the business here. Mr. R.H. Dixon is superintendent and has held this position for 8 years satisfactorily to all parties. E.C. Fields is foreman of the woodworking department.

H.N. Kennedy in the iron department, George Haisel-dean in the paint shop, J.F. Coddington- cutting, packing, and shipping department and Hugh Furgeson of the foundry; all of these men are thoroughly familiar with their respective branches, and have the confidence of their employees and the men in their divisions.

Mr. F.H. Lull is the secretary and book-keeper, which position he has very ably filled many years and to whom we are indebted for much information.

Considerable talent has been developed since the building of the manufactury here, in the inventing of improvements. J.F. Hollister, the oldest pattern maker here, has done much toward the improvements of the Marsh in its early history. Two of the foremen, Messrs. Coddington and Kennedy, invented a device for adjusting the reel; also an endless elevator, and other improvements, which have come into general use, and J.F. Steward, who is still in the employ of the company, is the patentee of a wire twister, which is used on the Crane binder, on which he receives a royalty. F.J. Coddington is also the inventor of patent chain elevator now used on the Marsh."

In addition to the sales agents' travels to many states and abroad, Gammon & Deering's products were sold in Plano by the firm of Henning and Ross. On Main Street near the train depot Henning & Ross had a large warehouse and sales room, built in 1877, the year they went into business.

Manufacturing for the 1878 season wrapped up in August; equipment and facility repairs were underway, and construction of a new 110-foot-long warehouse, directly east of the shops, on the east side of Hale Street, began. With E.H. Gammon living in Plano and William Deering visiting periodically from Chicago, the company's success continued though the business relationship of the two men was strained by Gammon's opposition to Deering's insistence on producing automatic binder attachments.

The factory instituted a ten-hour work day that fall. Production for the following season included machinery shipped as far as New Zealand for the southern hemisphere harvest season.

Gammon & Deering were faced with a local issue about this time. In March 1878, Israel Rodgers, Plano resident and bishop of the Reorganized Church of Jesus Christ of Latter Day Saints, threatened a suit against Gammon & Deering, or the town of Plano, or both, because the expansion of harvester factory buildings had closed off James Street near Rodgers' property which was located on James between John and Main Streets. Apparently as a result of the threatened suit, a petition was circulated in Plano in May and presented to the town's board of trustees. Signed by 160 business owners and taxpayers, the petition asked the town's leadership to consider extending to the harvester factory, "the territory they now need for enlarging and extending their works by vacating that portion of John Street laying East of Plain (author's note: Plain = Center Street/Avenue), and extending to the west line of Hale street in the town of Plano along the south side of the shops, ..." As noted in the *Plano Mirror* newspaper, "...all our business men with scarcely an exception, have signified their willingness and determination to encourage and assist our manufacturers in extending their work."

The problem was still unresolved in February 1879. The city of Plano proposed to Gammon & Deering that the company buy Israel Rodgers' property south of the harvester shops. The city stated that Rodgers had followed through with his threat to file a suit which would cost the town considerable expense and effort to defend. If Gammon & Deering would purchase Rodgers' property, the city was willing to vacate John Street south of the factory shops. The harvester factory then offered to buy the property.

Within a week the matter became more complicated. Another resident objected to the proposed purchase of the Rodgers property, pointing out that there were two other residences affected by the closing of the sections of John and James Streets occupied by harvester buildings.

There is no record of how the matter was eventually settled but James and John Streets did not reopen, Israel Rodgers was living in Sandwich, Illinois by 1880 and the city paid him $352 in 1880.

...scattering his brains along the track

Injuries were not uncommon at the harvester factories but fatal injuries were. Gammon & Deering employee Christian L. Johnson was killed in April 1878 when a train hit him as he exited the factory while carrying lumber which obstructed his view of the train. The train dragged him 300 feet, "terribly mangling his body and scattering his brains along the track." (*Kendall County Record*)

CHAPTER 6

...new-fangled, half-fledged contrivances, calculated
to delude the farmers by representations of wonder-working
powers no machine yet possesses.
(Pamphlet of C.H. McCormick & Bros.)

Praise for wire binders gave way to dissatisfaction as farmers discovered the drawbacks of wire as a binding material. The wire could catch in the moving parts of the machine. It had to be disposed of after it was removed from the bundles, which was a nuisance since it would neither burn nor rot. If wire got into the threshing machines or the millers' grinders it could damage the machines or end up mixed in with the grain. And if livestock ingested the wire they could die.

Inventors had been experimenting for years with other binding material. Straw, twine, and rope were tried but a knotter for any of these materials was a more complicated device than the mechanism for binding with wire. And wire was more readily available than a suitable twine.

The problem was finally solved in the late 1870s. Among the inventors striving to create a better binder was John F. Appleby. After several years of effort, in 1878 Appleby succeeded in building a device to automatically bind grain with twine. He had started his efforts before the Civil War from which time comes the story, whether accurate or apocryphal, of Appleby's inspiration for his first attempts to invent a twine knotting device. It was said that he saw a girl playing with a Boston Terrier puppy and a jump rope. The girl dropped the rope onto the puppy's head, at which he shook and backed away, the rope ending up in a knot on the ground. Appleby first made a knotting device from wood. It imitated the dog's twists and turns. He then made a knotter using iron.

His efforts were interrupted while he served in the Union Army. After the war, John Appleby learned from the progress made by other inventors working on the same problem. He devised a twine binder which was attached to a Marsh style harvester so that the grain was lifted onto the machine for automatic binding. "Mr. Appleby placed his needle beneath the binding receptacle, and his knotting devices above; he placed packing fingers each side of the needle, and so shaped the latter that when moved upward in the operation of binding the bundle of grain, it served as a stop to check the flow while the bundle was being bound. He placed discharge arms over the receptacle in such a manner that the bundle, when completed, would be ejected. The grain was not only packed into the receptacle, but while being bound was

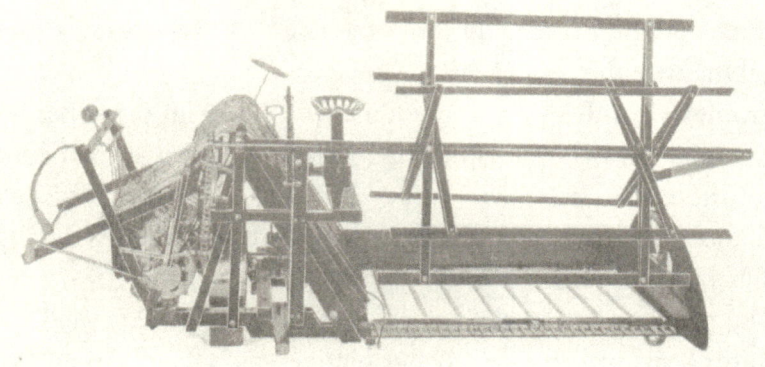

Appleby Grain Binder

suitably compressed, in order that a tight bundle might be the result." (*Deering Harvester Official Retrospective Exhibition, Development of Harvesting Machinery*, 1900)

Early in 1878 William Deering and John F. Steward were invited to Beloit, Wisconsin to see Appleby's twine binder, with the idea that Gammon & Deering might manufacture the device. Two of the binders were sent to Plano and from there to Texas where Steward tested them. Deering immediately recognized their potential. Gammon & Deering signed a contract in November 1878 for a license to manufacture the Appleby binders. Elijah Gammon opposed the risk involved in thrusting the new device on the market but Deering insisted. The first twine binders produced by Gammon & Deering for the harvest of 1879 did not work as well as hoped due to a lack of appropriate twine. They did, however, impress farmers, creating great demand for the binders. Deering spent months seeking a suitable and affordable twine, finally visiting rope manufacturer Edwin H. Fitler, of Philadelphia. Fitler's initial reluctance to adjust his machinery to produce the manila single fiber twine dissolved when Deering informed him that he would order ten rail cars of a twine that met his needs.

In April 1879, E.H. Gammon announced that he would retire that fall, selling out to William Deering. Gammon's health was the stated reason for his decision to retire, though his disagreements with Deering were likely a significant factor in the timing of his retirement. When he left the firm that fall he sold Deering his North Street home.

As the 1879 season closed in September, Plano's major business employed 400 men and produced machines for which there was great demand, including their new twine binders. These would eventually push aside wire binders but the wire binders were not yet obsolete as evidenced by 130 train cars of wire that arrived in Plano in May 1879 for use with the binders. The firm had shipped 8,000 harvesters, binders, and mowers for the season, amounting to sales of nearly $2,000,000. With the daring visionary William Deering at the helm, everything was looking up.

Farmers clamored for the twine binders, though the device had barely passed the experimental stage. Gammon & Deering had built 50 for 1879. The difficulty in finding suitable twine had left many manufacturers discouraged about the new devices but William Deering seized the opportunity, determined to manufacture 3,000 twine binders for 1880.

On Wednesday evening, September 24, 1879, the harvester factory engineer swept and straightened up and then left around 9:00. About 30 minutes later the night watchman passed through the boiler room. Then the superintendent came through. All was well, or so it seemed until shortly before 10:00 when the night watchman discovered flames leaping up in the boiler room from sawdust and fine shavings which had been piled in a corner to be used for fuel. The fire soon went up through a trap door to the story above. The watchman fought the fire but it spread quickly.

The Sandwich, Illinois fire department received an urgent telegraph message at 10:00 that evening. They were needed in Plano where the harvester factory was ablaze. Within 35 minutes the Sandwich firefighters arrived to find flames sweeping furiously eastward through the factory despite the efforts of employees and Aurora firefighters already at the scene. By the time the blaze was out, the total loss reached $35,000, including $3,000 worth of equipment purchased just two weeks earlier.

Several departments were heavily damaged, others not at all. Through the efforts of employees and townspeople the fire was confined to one building and lumber piles were saved. William Deering arrived Thursday morning to inspect the damage. The factory was not fully insured but Deering stated his intentions to begin rebuilding within two weeks.

Repairs to the harvester shop and its machinery were complete by early November. Large quantities of materials for the 1880 machines were delivered and Plano homes spruced up in expectation of prosperous times ahead. William Deering suggested he would build 1,000 more harvesters for 1880 than originally planned if the capacity of the Plano factory could handle the increase. Deering presented factory employees turkeys for Thanksgiving and sold them coal at an excellent price, making for cheerful holidays.

William Deering

William Deering
1826-1913
William Deering spent his early life in Maine where he began to
study medicine, but left to assist his father who managed a woolen
mill. Deering never returned to medicine. Instead he amassed a fortune in the woolens and dry goods businesses.

In 1870 he looked westward for investments. Meeting an old
friend, Elijah H. Gammon, while visiting Chicago, Deering found his
opportunity. He moved to Illinois in 1873 to become directly involved in the harvester manufacturing company from which he
would earn his second fortune.

After buying out Gammon, Deering moved the company from
Plano to Chicago where the success of the Deering Harvester Com-
pany led to its merger with the McCormick Harvesting Machine
Company and three other firms to form International Harvester
Company in 1902.

William Deering's interest in education became the focus of his
philanthropy. He contributed generously to Northwestern Univer-
sity and Garrett Biblical Institute, in Evanston, Illinois.

After his retirement in 1901, William Deering lived in Florida
where he died in 1913.

The factory was busy in early 1880. William Deering's son,
James, took on the job of assistant superintendent. With Gammon's departure, the elder Deering frequented Plano to oversee
the booming business. The factory was running night shifts again
with machinists coming from Chicago for the late shift. Machines
were sent across the United States, as well as to Europe, Australia,
and New Zealand. Two hundred twenty-five additional machines
were ordered from New Zealand in June, along with orders for
240 machines for the following year. Six to ten train cars of equipment were shipping out every day—the factory couldn't keep up
with the orders.

William Deering sold the home on North Street to Albert H.
Sears for $2,000 in July 1880. Sears, the son of one of Plano's

founders, Archibald Sears, was himself involved in harvester manufacturing.

Albert H. Sears Home on North Street, Plano

Albert Hollister Sears

> Albert Hollister Sears
> 1856-1917
> Albert H. Sears was the son of Archibald Sears and Rachel Carver Smith who were among the founders of Plano, Illinois. Sears spent much of his childhood living with his aunt and uncle, Emeline and John Hollister. He worked for the Deering Harvester Company in Plano and was one of the founders of the Plano Manufacturing Company. Sears later operated a hardware store and a bank in Plano. After the Plano Manufacturing Company moved to Chicago, Sears bought the vacant factory buildings and operated Sears Manufacturing Company.
> Albert H. Sears purchased the house of William Deering (built by E.H. Gammon) on North Street in Plano in 1880 and expanded and transformed it into a beautiful Queen Anne style home which is now on the National Register of Historic Places. Sears also owned the summer resort known as Millhurst on the Fox River south of Plano.

Plano was booming, but troubling talk floated among the sawdust in the harvester shops. Rumors of Deering's intention to build a factory in Chicago concerned the workers. Some were certain he wouldn't close the Plano facility even if he built new shops elsewhere. The many carloads of lumber surely indicated his intention to continue manufacturing in Plano.

Then, the *Kendall County Record* reprinted an article from the July 11 edition of the *Chicago Tribune*.

"The Marsh Harvester Company has broken ground for their works, corner of Fullerton Avenue and the Chicago and Northwestern Railroad. They have put down about eighty rods of rail and contracted for a large number of brick. Their purchase includes 25 acres and will all be occupied by their works and the cottages they will build for their workmen. They will employ about 300 operatives and will add something like 1,000 to the population of the district."

William Deering recognized the limitations of the Plano factory—both its size and location. He expected he'd recoup the $50,000 construction cost of the Chicago factory in two years' time from reduced shipping costs. Despite inducements by the town, including an offer of any land in Plano, Deering had decided to leave Plano behind.

Deering Marsh Harvester with Twine Binder, 1880

CHAPTER 7

In the ten years leading up to 1880, Plano's population greatly increased, due almost entirely to its harvester factory. But in 1880, many of Deering Manufacturing's employees followed the company to Chicago. Some who remained in Plano were unemployed. If the empty Plano factory were to stand idle for long it would be devastating for the town.

The residents of Plano sprang into action that summer. Over 200 townspeople gathered at Good Templars Hall on West Main Street on August 21st to discuss organizing a joint stock company. William T. Henning led the meeting. Lewis Steward spoke at length about the need for a new manufacturing company to occupy the harvester shops. He advocated the involvement of the entire town. Steward proclaimed his confidence in Plano's mechanics and laborers and put forth the idea that many Plano residents could afford to buy at least one share in a stock company.

Others added that a company owned by the townspeople would be less likely to move away due to the decision of one owner. The halt in public improvements caused by the departure of Deering and his firm was a concern. By a unanimous vote the decision was made to organize a joint stock company.

Two weeks later, business people, mechanics, and farmers met again at Templars Hall where the subscription books for stock purchases were opened amid enthusiasm and expressions of support. Mechanics were early subscribers in the new company. A few of the wealthy subscribers had already sweetened the pot and planned to purchase more stock as necessary.

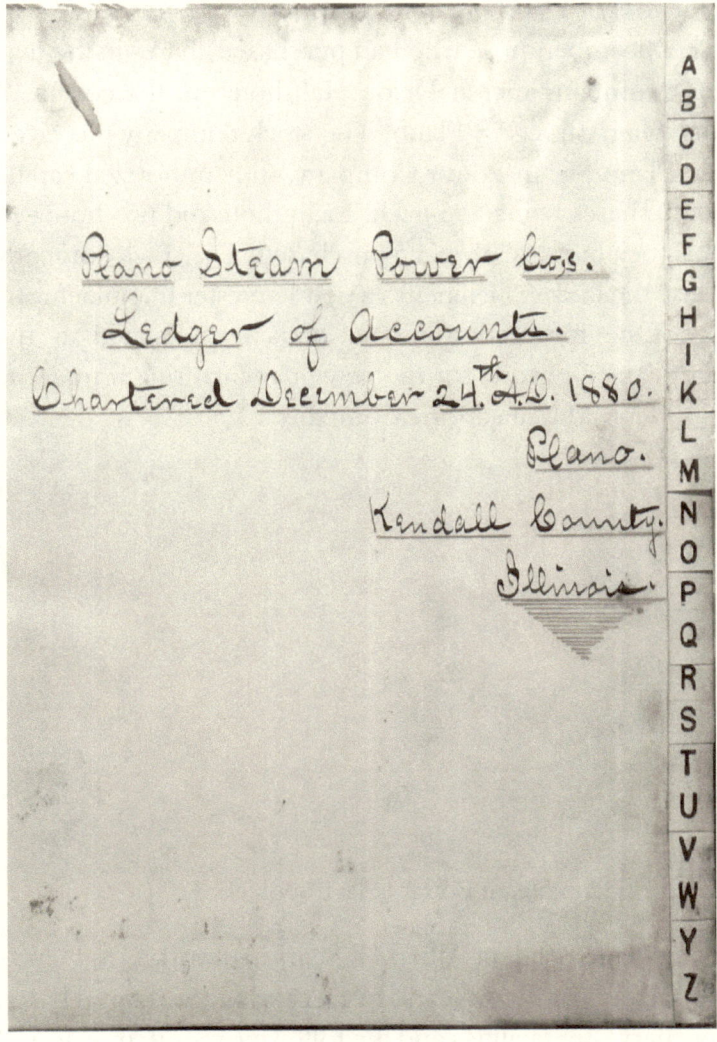

November 13, 1880 was announced as the permanent closing of Deering's Plano factory. Other manufacturers had visited the town to consider the soon-to-be-empty facilities but none made a move. An assessment of the town's financial status published in December showed that neither the cost of settling with Bishop Rodgers over his suit regarding the closing of James Street nor expenditures related to a smallpox outbreak earlier in the year were substantial. The report would have seemed encouraging if not for the loss of Plano's major enterprise.

In late December those who had purchased shares in the new joint stock company met at Dixon Hall hotel on the corner of West and Main Streets in Plano. The stock company was given the name Plano Steam Power Company, with authorized capital of $10,000, shares being $25 each. Eight thousand two hundred dollars of stock had already been purchased. The stock company would lease the factory facilities to a new harvester manufacturing company. One hundred thousand dollars was stated as the amount of capital needed for the new manufacturing firm, to be known as Plano Manufacturing Company.

Plano Steam Power ledger book entries

Lifelong Plano resident Albert H. Sears had worked in Plano's harvester factory for many years. In December, Sears purchased the empty harvester facilities and the following month transferred the factory to the Plano Steam Power Company which then leased it to the Plano Manufacturing Company. On January 20,

1881, Lewis Steward, president of the Plano Steam Power Company, and Albert Sears, its superintendent, met with William Deering in Chicago where they obtained the title to the Plano harvester works.

Elijah H. Gammon, Deering's former partner, was a large stockholder in the Plano Steam Power Company as was W.H. Jones, a longtime sales agent for Deering's Minnesota territory. Jones was now ready to join the leadership of Plano Manufacturing Company. With the $100,000 capital stock of Plano Manufacturing Company all sold, a board of directors and officers were chosen. W.H. Jones would be President, Lewis Steward, Vice-President, G.W. Chamberlain, Secretary, and Albert H. Sears, Superintendent. These men, along with Gammon's son-in-law, Dr. Fredrick J. Huse, made up the company's first board of directors.

Lewis Steward

Lewis Steward
1824-1896

Marcus Aurelius Steward brought his family from New England to Wayne County, Pennsylvania, and then in 1838, to Kendall County, Illinois where he bought a large tract of land in what is now Plano. Lewis, the oldest of Steward's seven children, worked on the family farm and at their sawmill. He studied law and joined the Illinois bar though he never practiced law.

In 1853, Steward persuaded the Chicago, Burlington, & Quincy Railroad to direct their westward extension through his family's land, promising to establish a town. The town of Plano was laid out that year.

Steward owned Plano's first grain elevator, its first bank, a tannery, a boot and shoe factory, and a flour mill. He built the town's first water system, opened an opera house, served as president of two railroads, gave land to establish Plano's first churches, provided gravel for roads, owned 4,000 acres of active farmland, and helped establish the Marsh harvester factory in Plano. He accomplished so much that Plano was sometimes referred to as "the child of Lewis Steward's creation".

In 1876, though not a typical politician, Steward ran for Governor of Illinois, failing to carry Kendall County but only narrowly losing the election. He was elected to the United States House of Representatives as a Democrat in 1890, serving one term.

Steward and his second wife had seven children. His sons were named after friends or business associates, including Julian Rumsey Steward, named for Julian Rumsey, founding member of the Chicago Board of Trade and mayor of Chicago; William Deering Steward (see William Deering biographical sketch on page 55); George Somerset Bangs Steward, named for George Somerset Bangs, General Superintendent of the Railway Mail Service; and Charles Marsh Steward (see Charles Marsh biographical sketch on page 28.)

Steward died in Plano in 1896 at the age of 71. His 1854 house on East Main Street in Plano is listed on the National Register of Historic Places.

Plano Manufacturing Company expanded the manufacturing plant and by March 1881 was ready to build harvesters. It was a late start for the harvest season and they were forced to construct many of the harvester parts themselves. But their first sample machine was in the able hands of experienced mechanic and inventor, John F. Hollister.

The new harvester, which included Hollister's own improvements, was completed by early June. After a trial in a field of tangled rye at A.E. Henning's farm, the harvester's unusually light draft was proclaimed "the lightest he had ever tried" by Mr. Rounds who operated the machine. The new harvester featured... "that the pitman is attached directly to the end of the sickle as in the case of mowers, instead of the attachment in the center with a reverse action from the rear as in most other harvesters, thus avoiding much friction and wear. It has a canvas elevator, and the gearing is run by the endless chain." (*Kendall County Record,* June 9, 1881). The company guaranteed that this Jones Lever Binder, as the 1881 machine was called, was the lightest running binder on the market. It was available in five, six, or seven foot sizes.

1881— Plano Manufacturing Company employees were male with the exception of several women who worked in the canvas room operating sewing machines.

With the success of the sample self-binding harvester, Plano Manufacturing began operating on a twelve hour per day construction schedule in order to build as many harvesters as they could manage. In addition to the harvesters, W.H. Jones had secured the exclusive right to build the Warrior mower whose construction would begin after the harvesters were complete.

By the end of summer 1881, two hundred fifty harvesters had been built and shipped, most headed to Minnesota, Iowa, and the Dakota Territory. Construction of mowers began in September. The shop office was now complete, with flooring of black walnut and cherry and two fireproof vaults.

For 1882, construction of 1,500 twine binder harvesters and 500 mowers was the goal for the new company. W.H. Jones moved from Minnesota to Plano to directly oversee the business and Elijah Fields, formerly with the Deering Company, returned to Plano to replace Albert H. Sears as superintendent. Sears had decided to open a bank and mercantile business on the northeast corner of Main and Hugh Streets in Plano, but retained his financial interest in the Plano Manufacturing Company.

The new harvester firm, so crucial to the town of Plano, was well supported by the townspeople. When the company was formed, a group of Plano's women collected funds which they used to purchase a chime whistle for the firm to signal meal breaks and the end of each work day. They also bought a brass locomotive clock for the factory's engine room. The gifts were formally presented at a banquet at the factory attended by 500 Plano residents, during which the whistle was demonstrated. The demonstration, held at a time of day when the whistle would not normally chime, drew the attention of the Bristol and Sandwich fire departments who feared another blaze in Plano.

February 11, 1882 was a day of mourning in Plano. John Fletcher Hollister who had helped the Marsh brothers create the first successful harvester and had designed Plano Manufacturing's first harvester, died at age 71.

In June 1882 Plano Manufacturing doubled their capital stock to $200,000. Forty thousand dollars of the original stock was owned by Elijah H. Gammon and Dr. Fred Huse, both of Batavia. The additional capital was to enable the company to build 8,000 machines for the 1883 season. Demand for the Plano twine binders was rapidly increasing. Some orders had to be turned down because the factory could not manufacture enough.

Increased production required more employees which caused a housing shortage in Plano.

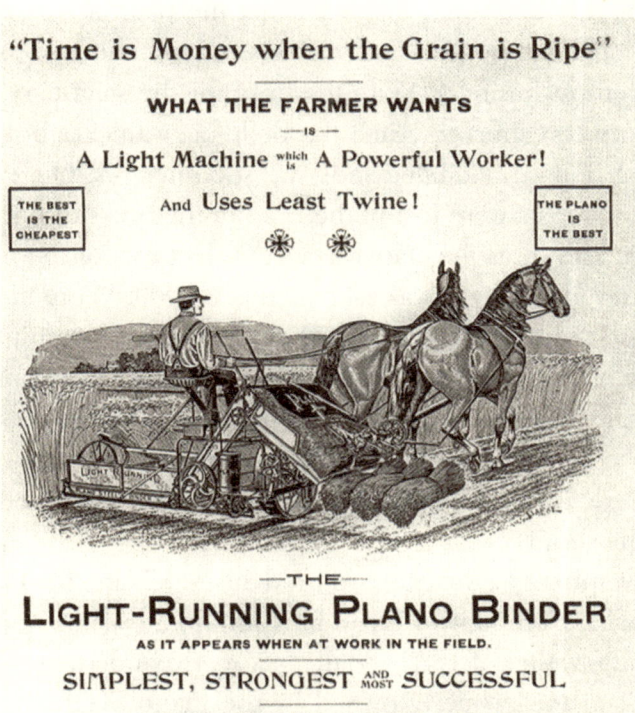

"Time is Money when the Grain is Ripe"

WHAT THE FARMER WANTS

— IS —

A Light Machine which is A Powerful Worker!

And Uses Least Twine!

THE BEST IS THE CHEAPEST

THE PLANO IS THE BEST

—THE—

LIGHT-RUNNING PLANO BINDER

AS IT APPEARS WHEN AT WORK IN THE FIELD.

SIMPLEST, STRONGEST AND MOST SUCCESSFUL

LONGEST LIVED AND LIGHTEST DRAFT.

The Only Machine Pulled EASILY by Two Horses!

Another Fire

Plano Manufacturing's whistle blew unexpectedly on the morning of July 20, 1882. Its signal that the paint shops were ablaze brought employees and townspeople, anxious to save Plano's foremost business. With the hose for the stationary fire engine pump too short to reach the building, a bucket brigade was formed. The smoke-filled shop was soon inaccessible. Paint and other materials were lost in the fire but the one-story frame store house known as the "tabernacle", 60 feet east of the paint shop, was saved. Damage was estimated at $5,000. Women and children had helped with the water buckets, with the result that the women realized the need for a longer fire hose. They immediately began collecting funds for that purpose. Their efforts resulted in the October 1882 installation of a water pipe and hydrant just south of the harvester shops, extending the entire length of the shops.

In November, Lewis Steward sold his stock in the Plano Manufacturing Company to Elijah H. Gammon.

During the winter of 1882—83 the factory expanded, purchasing the rest of the property owned by the Plano Steam Power Company. More modern equipment to speed construction of the machinery was also installed. As 1883 progressed, Plano Manufacturing moved homes from the newly acquired property and constructed additional buildings. The town happily accommodated their largest employer, granting them permission to occupy parts of John Street, James Street, and Hale Street indefinitely.

...simply to gratify some hair brained lunatic of an inventor...

For 1883, Plano Manufacturing Company advertised:
"Buy the machine that has the least possible machinery and the least number of pieces. The fewer the better. Don't buy a complicated trap that is fearfully and wonderfully made, whose chief utility and advantage

consists only in talking points for windy agents. You probably are not an expert in machinery and life is too short for a farmer to spend time to learn the machinist's trade simply to gratify some hair brained lunatic of an inventor who has experiments he wishes you to buy and try. Let them alone...Let the horse killers go."

A NEW INVENTION!

STORED POWER SUCCESSFULLY APPLIED
TO A HARVESTER AND BINDER!

THE
FLY-WHEEL
ATTACHMENT.

IF A BINDER FAILS ANYWHERE IT IS IN SOFT, WET GROUND AND IN HEAVY, TANGLED GRAIN. WITH OUR FLY-WHEEL ATTACHMENT THE PLANO CANNOT FAIL.

A FEW OF THE

MANY REASONS

WHY

THE PLANO FLY-WHEEL

IS A

VALUABLE APPLICATION

TO A HARVESTER

AND BINDER.

Plano Binder with Flywheel attached

Plano Manufacturing Company advertised the features
of its 1883 Light Running Plano Twine Binder:

1. An adjustable trip to make any sized bundle desired. The ad contrasted it with other binders designed to trip with packers.

2. Simple arrangement of raising and lowering, contrasted with binders that raised with levers, ratchets and chains running under the platform which the ad claimed added 150 to 200 pounds to the competitors' machines.

3. The reel worked with one lever and foot trip and could be stopped when the machine was thrown out of gear.

4. Plano's twine binder had a cold rolled angle iron bar instead of a common wrought iron and wood bar as used by many other manufacturers. The cold rolled bar was claimed not to warp or get out of shape. The shafts were also made of cold rolled iron.

5. The sickle had direct connection with the pitman (connecting rod) with which the knives could be taken out easily, instead of what the ad described as an "old fashioned rattle trap sway bar".

6. It had fewer parts than other binders, making it less complicated.

Machines for sale for the 1883 harvest
PRICES TO SALES AGENTS:

Plano Harvesters	6½ foot cut	With Twine Binder	$280.00
Plano Harvesters	6½ foot cut	With Hand-Binding Tables	$175.00
Plano Harvesters	5½ foot cut	With Twine Binder	$275.00
Plano Harvesters	5½ foot cut	With Hand-Binding Tables	$170.00
Binding Twine			20¢/pound
New Warrior Mowers	4 ft 3 in cut		$75.00
New Warrior Mowers	4 ft 8 in cut		$80.00

The Light-Running Plano Harvester and Twine Binder

Light-Running Plano Harvester and Twine Binder, rear view

...a bitter rivalry...

Elijah H. Gammon owned $100,000 of stock in the Plano Manufacturing Company and had lent the enterprise an additional $200,000 as of 1884. The *Sycamore True Republican* newspaper cited a "bitter rivalry between him and Deering, his old partner..." with each striving to outdo the other in the extent of their business.

Plano Manufacturing Company moved its business office from Plano to Chicago in the fall of 1885. This sparked speculation that the entire plant might soon move also, but company president W.H. Jones asserted that the firm had no such intentions.

Concerns arose in Plano again the next year that its main enterprise might depart for lack of sufficient land in which to expand. Many residents remembered the dark days surrounding the departure of The Deering Manufacturing Company less than ten years earlier. The citizens of Plano had cause for much relief in the summer of 1887 when city officials, certain landowners, and Plano Manufacturing Company were able to come to an agreement that provided the manufacturing firm with the land it needed.

PLANO MANUFACTURING WORKS.

Plano Manufacturing's all-steel harvesters for 1888 were a success, the firm receiving more orders than they could fill. The company replaced the harvesters' gears with a chain drive in the 1888 models, which they advertised as the first harvesters to use the chain drive.

Their 1889 advertising noted U.S. distribution points of eighteen cities spanning fifteen states plus the Dakota territory. The company also maintained three foreign offices: Milan, Italy; Paris, France; and Buenos Aires, Argentina. The company continued to prosper. Twenty thousand machines were produced in 1890. Following the success of their chain-drive harvester, they now offered both a chain-drive mower and a geared mower.

An attempt to form a harvester monopoly

Intense competition among the various producers of agricultural equipment forced prices downward over the last two decades of the nineteenth century. Plano Manufacturing's 1891 advertisements informed readers of the formation and quick demise of the American Harvester Company which was essentially an attempt by six large harvester manufacturers to combine to form a monopoly, including buying a number of smaller companies. McCormick Harvesting Machine Company and William Deering and Company were the two largest firms involved. Although its 1891 advertisement claimed Plano Manufacturing Company was not involved in the merger, company president W.H. Jones had informed employees in 1890 that they would become American Harvester Company employees *(Kendall County News,* December 11, 1890). The American Harvester Company plan quickly fell apart.

1893 Jones Chain Power Mower

By 1892 the chain-drive mower, known as "the Jones" was considered a success. The chain drive self-binding harvesters had long been successful such that other manufacturers had been convinced to adopt the chain drive for their own harvesters. "The principle of transmitting power directly from the driving wheel, or its axle, of a harvesting machine by a chain or link belt is now universally acknowledged to be the best and most effective" proclaimed the May 5, 1892 *Farm Implement News.*

By this time, Plano Manufacturing was doing a booming business and paying out large dividends. In 1888, Plano Manufacturing was recognized as, "third in the list of producers of "binders and mowers" having a working capacity of 12,000 binders and 5,000 mowers annually, and giving employment to 500 men; their average monthly pay roll figures from $25,000 to $30,000. *(Portrait and Biographical Record of Kane and Kendall Counties, 1888)*

For the 1891 season, the Plano harvester/binder featured a flywheel attachment,

> "...*by which all strain upon the machine and extra draft upon the horse—while the bundle is being compressed and bound—are avoided, absolute steadiness of motion is obtained, liability to breakage when striking obstructions is removed, capacity and durability are increased and better work and complete adaptability to all the unfavorable conditions in harvesting are accomplished. This fly-wheel is connected with the main gear shaft in the rear of harvester and driven through a spring clutch device ....*" (The Farm Implement News, January 19, 1893)

Readers of the *Chicago Tribune* opened their newspapers on January 29, 1893 to the news that the Plano Manufacturing Company had purchased a twenty-five acre site in Chicago as the new location for their factory. The new site was located along the line of the West Pullman branch of the Illinois Central railway. In 1892, Plano Manufacturing's 600 employees had built over 43,000 machines at its Plano facilities, a plant which included twenty buildings.

The company had outgrown the Plano site and the railroad facilities at Plano were no longer sufficient for the growing enterprise. Plano Manufacturing Company, originally promoted to Plano's residents as a company owned by the townspeople and, therefore, less likely to relocate elsewhere, was now controlled by a small number of men who no longer lived in Plano: W.H. Jones (President), his brother, O.W. Jones (Secretary), L.B. Wood (Treasurer), and J.P Prindle (Vice-President and successor to his late brother-in-law, Elijah H. Gammon).

The late January news was not a complete surprise to Plano residents. It had already been reported that concerns about insufficient railroad facilities had led Lewis Steward, then a United

States congressman, to explore the establishment of a new railroad connecting Plano and DeKalb. Steward had met with the Ellwoods, barbed wire manufacturers in DeKalb, and discussions had included the possibility of connecting a railroad from DeKalb through Plano to Joliet. Ultimately, however, there was no new railroad through Plano.

Ground was broken for Plano Manufacturing's new factory in Chicago's West Pullman neighborhood in May 1893 on the day the World's Columbian Exposition opened in Chicago. That

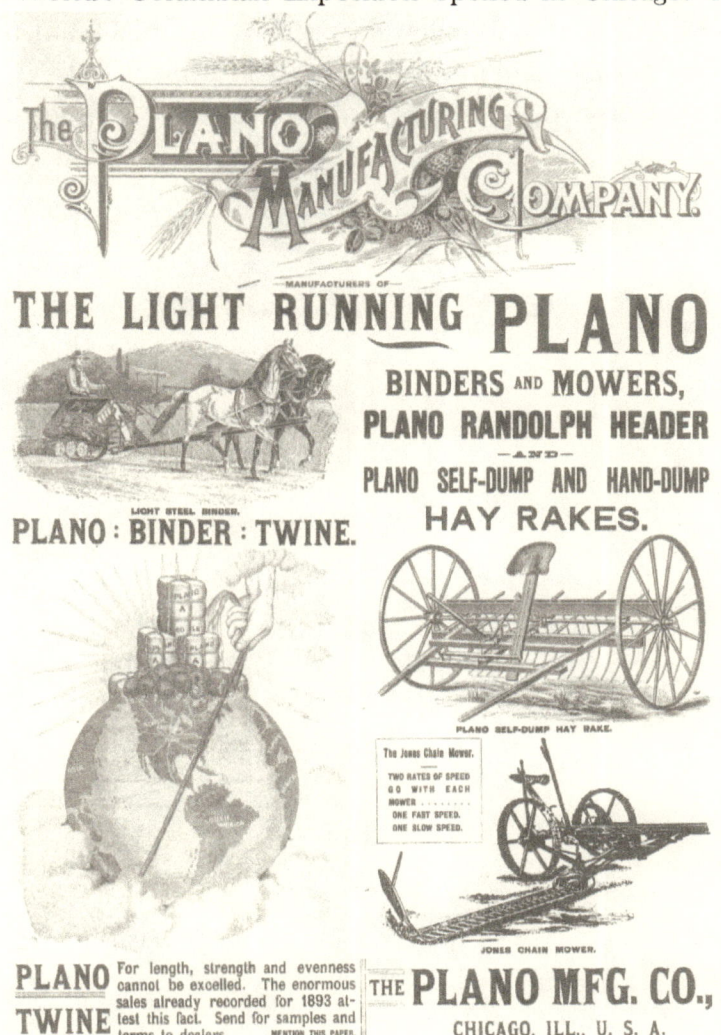

same week a severe financial panic began in the United States, lasting from May to November 1893. It was followed by an economic depression that lasted until 1897. Many banks failed, the stock market declined and unemployment was high.

In Plano, the factory was still operating over the summer of 1893 while construction of the new facilities in Chicago continued, with September 1 as the target date for completion. Some Plano residents were in denial about the impending departure of the factory but Plano Manufacturing Company's strong financial status enabled the firm to follow through with the establishment of the new facilities and to weather the 1893 panic and the depression that ensued.

CHAPTER 8

Even before Plano Manufacturing Company's groundbreaking for the West Pullman factory, rumors were flying in Plano about plans for a new company to occupy Plano's soon-to-be empty shop buildings. Losing the town's main employer could not have come at a worse time. The economic crisis dampened hope of another large enterprise arriving to occupy the 20 empty factory buildings in the center of town.

In 1895 some of the deserted factory buildings found new occupants. The Plano Implement Company was formed by local men J.D. Morris, W.W. Owen, F. W. Lord, C.M. Morris, A.C. Sanders and J.E. Turpin. With capital stock of $38,000 the firm manufactured the Davenport Potato Cutter and Planter, which, according to its advertising "...marks, furrows, drops and covers all in one operation. Does away with cutting seed by hand only one piece in a hill, never misses, no seed wasted. Cuts the potatoes if done by hand and leaves the field with its work completed. Thoroughly tried and successful. Won first prize in field contest at Iowa State Fair in 1895."

Although the firm did not manufacture harvesters, farmers could take their harvesters to the Plano Implement Company for repairs.

Albert H. Sears, whose bank in Plano was said to be the only Kendall County bank to continue successfully through the economic panic of 1893, bought the Plano Implement Company in 1897. He operated it as Sears Manufacturing Works, continuing to build potato planters, along with lever harrows, disc harrows, corn planters, corn cultivators, manure spreaders, brass and iron beds, ladders, and lawn furniture.

SEARS SURFACE CULTIVATOR.

Has No Equal.

Easiest to operate.   Most adjustable.   Throws dirt in or out.
Perfect balance.   Break pins on shovel standards.
Works in any soil.   Simple, Light and Strong.

NO COMPETITION FOR THE DEALERS WHO SELL IT.

Manufactured by

ALBERT H. SEARS,

PLANO, ILLINOIS.

SEND FOR CATALOGUE.   Mention this paper.

In the late 1890s John F. Steward began writing his account of the invention of the harvester and subsequent innovations. Steward wished to record what he considered to be accurate documentation of the history of the harvester. His ill feelings toward the McCormicks, longtime rivals of Plano's harvester factories, were evident in much of what he wrote of the invention and manufacture of reapers and harvesters. Steward corresponded with Charles W. Marsh and many others, soliciting their recollections about the history of harvester inventions and manufacturing.

In his posthumously published book, *The Reaper, A History of the efforts of those who justly may be said to have made bread cheap*, Steward recounts the 1897 attempt of the McCormick company to have a portrait of Cyrus H. McCormick used on the United States Treasury Department's ten dollar silver certificate. Several harvester manufacturers, and at least one periodical protested the plan, noting that McCormick's likeness was used in advertising and members of McCormick's own family claimed his father was the actual inventor. The Treasury Department reversed course and did not use McCormick's picture.

SEARS & CO.,
PLANO, ILLINOIS,

Have the agency for Kendall county, and have just received a car load of those famous Miller Spreaders, and would be pleased to show you how the Spreader is of as much value to any farmer as any tool he uses on the farm.

We would be pleased to hand you list of farmers who purchased the Miller Spreader last season and accept their recommend.

Some of the Strong Features of the Miller Spreader:

1st.—It is very low down, making it easy to load.

2nd.—It is just the right width to straddle corn rows and leave all the land evenly covered with manure. This a strong feature on corn stubble land, owing to the fact that other machines cannot do good work without driving team and Spreader wheels on the cut corn hills, a thing that farmers know is impossible.

3rd.—The Miller is the only Spreader that will do perfect work in winter with manure from the stable daily. Why? Because we have a solid bottom which is scraped clean every load. Nothing can freeze to it as in tread power machines.

4th.—Being low down, it enables us to get in and out of sheds where others fail.

5th.—We can start or stop cylinder independent of feed. Either is operated at will of driver.

6th.—We can put on 6, 10, 15 or 20 loads per acre. and unload in from four to six minutes.

7th.—To top dress a meadow or pasture it has no equal.

8th.—It doubles the value of manure by covering two acres where you could only cover one by hand.

9th.—It is simple,strong and durable, having been on the market fourteen years and is no experiment.

We also desire to call your attention to the fact that we carry the largest line of Surreys, Buggies, Wagons, Milk Wagons and Farm Machinery in Kendall county. Our prices are as low as any of our competitors, and we handle direct from the Manufacturer and guarantee every line we sell.

See our large line of Double Driving, Single and Team Harnesses. One year's guarantee on every Harness sold.

We also have a complete and full line of Steel Ranges, Stoves and Hardware, Lawn Mowers, etc.

It is no trouble for us to show goods, so kindly call in and get our prices

SEARS & CO.,
PLANO,   I   I   I   ILLINOIS.

In 1908 Sears incorporated his company, keeping the name, Sears Manufacturing Works. Capital stock was $250,000. Along with Sears, the incorporators were his brother, James M. Sears, and Charles A. Darnell. At that time the firm had 75 employees.

Several other factories operated in Plano between 1898 and 1908. Most of the businesses failed within two years. Some used at least one of the buildings that had been part of Plano Manufacturing Company. One business that lasted several years was Earl Manufacturing Company, maker of lawn swings, ladders, and wheel barrows. It occupied buildings just east of Sears Manufacturing Works.

"The name of Plano is worth millions of dollars
to any harvester concern."
—Elijah H. Gammon

A harvester manufacturing firm returned to Plano in 1905 with the arrival of the Kellogg Harvester Company. Company president E.M. Kellogg, invented the Kellogg Packerless Grain Binder and the Kellogg Packerless Corn Harvester and Binder, which the company planned to manufacture in Plano. Kellogg's corn harvester was unique in that it bound the corn as the corn leaned forward, coming from the cutting knives. In contrast, the Deering style corn harvester bound the corn lying down and the McCormick style harvester bound the corn by standing it upright. Kellogg's was called "packerless" because it did not need mechanical packers which he claimed knocked down the ears.

After working for various large manufacturers, Kellogg had founded his harvester manufacturing company in Chicago about a year earlier. The vice-president, treasurer and manager of his company was William Campbell Thompson. Kellogg and Thompson sold stock in their company throughout 1904. In a November visit to Plano, Thompson stated that he felt confident they would raise the $25,000 "required to be raised here among

Plano people for the Kellogg Harvester Co." (*Kendall County News*, November 16, 1904). $100 shares were offered to Plano residents, payable in four monthly installments.

By January 1905 Kellogg Harvester had raised $25,000 from Plano residents and considerable funds in Chicago. Having sold his Chicago bond and brokerage firm, Thompson ran the business end of Kellogg Harvester Company while E.M. Kellogg and his sons would be in charge of the mechanical department.

Kellogg Harvester planned to purchase the 6,000 square foot factory building along the railroad from the F.H. Earl Ladder Company, just east of Sears Manufacturing. They hoped to add a large machine shop and an additional foundry for the manufacture of grain binders, corn binders, mowers, shockers, and cultivators. The *Kendall County News* announced the plans as a certainty in January but by mid-February the deal had fallen through.

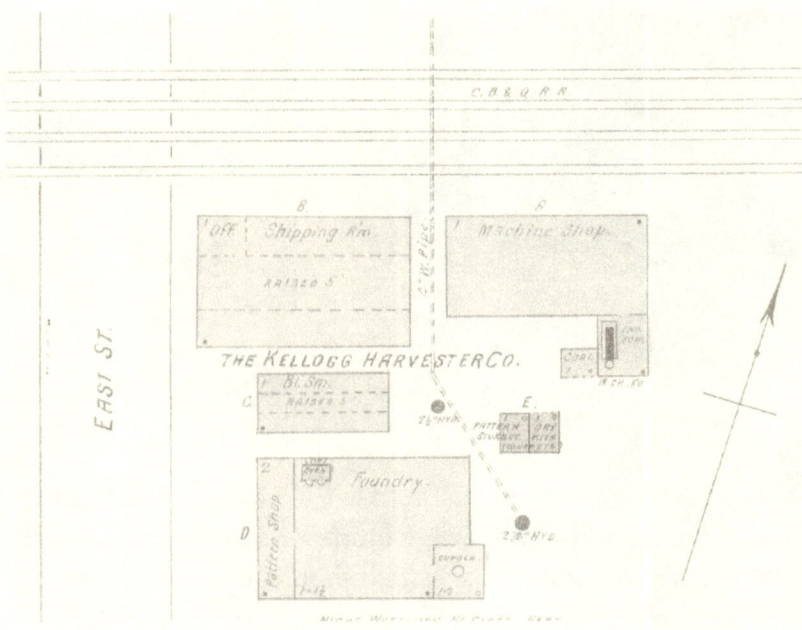

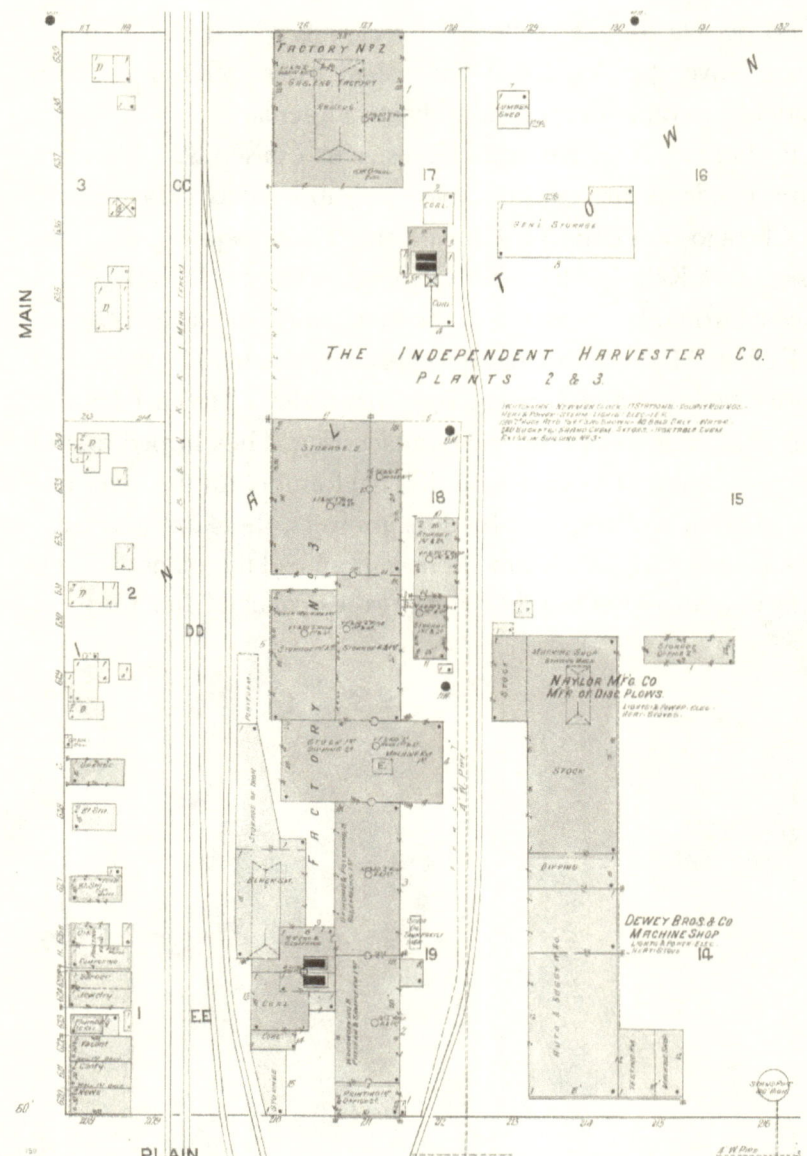

Instead, Kellogg Harvester purchased the factory building of the Long brothers' woodworking company, east of F.H. Earl's factory. A machine shop, a blacksmith shop, and a foundry were then constructed, with plans to begin building their products by the beginning of May. On the heels of this report in the *Kendall County News* came a report in mid-June that Kellogg Harvester

84

had purchased ten acres of land from A.H. Sears, which was described as extending from Sears's factory westward, providing 1,420 feet of rail switch fronting. This was where their new and permanent buildings would be located.

Many company employees were stockholders. Thompson and his staff peddled the stock energetically, resulting in stockholders in five states by summer 1905. Inventor Joseph Boda, the factory's General Superintendent, was elected a company director, joining directors Julian R. Steward, E. M Kellogg, Ning Eley, William C. Thompson, and Donald Fraser (Thompson's father-in-law). In November 1905 the Kellogg Harvester Company changed its name to the Independent Harvester Company.

Independent Harvester was still getting on its feet. The company shipped its first products—manure spreaders and corn harvesters—to Iowa. Inventor William Hibbs moved from Iowa to Plano to join the company and continue working on his corn picker and harvester. Around this time E. M. Kellogg left Independent Harvester.

William Hibbs with his corn picker and harvester

According to company president William C. Thompson, 1906 to 1908 were spent setting up the company, acquiring property, building a foundry, and blacksmith shop. An "experimental farm" was purchased in Plano near gravel pits along the Big Rock Creek and including the home at 405 S. Hale Street.

In November 1909, Albert H. Sears retired from the manufacturing business, but continued to operate his bank on Main Street in Plano. He sold the A. H. Sears Manufacturing Company building and property to the Independent Harvester Company, including the patents, patterns, machinery, and good will, for $40,000.

Independent Harvester's purchase of Albert H. Sears's factory meant they now owned all of the grounds and buildings that had belonged to the Plano Manufacturing Company. Independent Harvester had purchased the Earl Ladder factory a few weeks before their acquisition of the Sears plant. Sometime before the sale, the Earl factory had burned. Independent Harvester replaced it with a new building they called Factory No. 2 in which they planned to manufacture plows, harrows, and other farm implements. Factory No. 1 was to the east and Factory No. 3 was the original harvester factory building on Plain/Center Avenue.

Company offices were located at the factory until 1909 when the company bought the former residence of Dr. Daniel Jenks, originally the home of George S. Steward, at 307 E. Main Street in Plano.

INDEPENDENT HARVESTER OFFICE   PLANO ILL.

Independent Harvester Co., Factory No. 1, Plano, Ill.

Independent Harvester Factory No. 3, view from Plain/Center Street

In contrast to other companies, Independent Harvester sold its products directly to farmers, bypassing dealers. Its entry into grain harvester manufacturing was slow; its products in 1909 were manure spreaders, grain shockers, gasoline engines and corn harvesting machines. Much stock was sold with the promise to farmers of a discount on the purchase of a binder—whenever the company actually began manufacturing them.

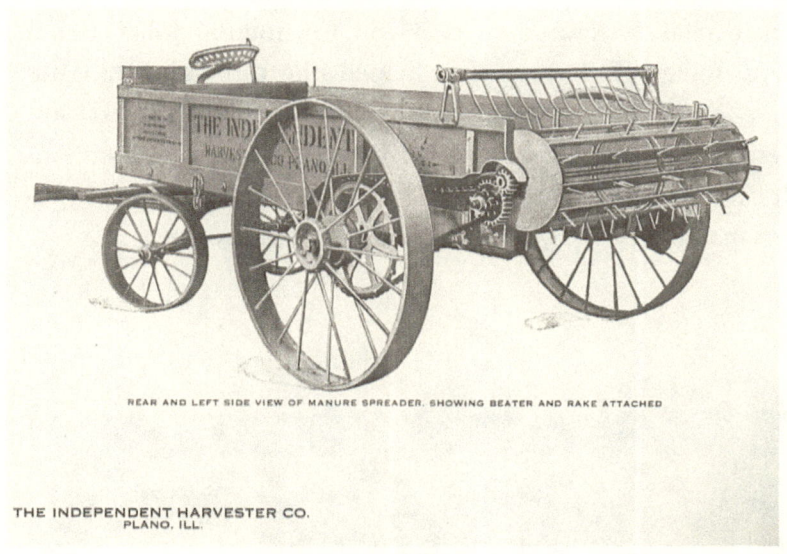

REAR AND LEFT SIDE VIEW OF MANURE SPREADER, SHOWING BEATER AND RAKE ATTACHED

THE INDEPENDENT HARVESTER CO.
PLANO, ILL.

Independent Grain Binder

The Independent Harvester Company was organized as a co-operative. A holder of a share of preferred machinery discount stock was entitled to buy farming implements and parts at a discount. At the company's 1909 annual meeting of the stockholders the capital stock was increased from one million dollars to ten million dollars. The stock was divided into three classes: common stock (28,000 shares), preferred stock (2,000 shares), and preferred machinery discount stock (70,000 shares). Shares of each class of stock were sold for $100 each. Holders of the preferred machinery discount stock had no voting power.

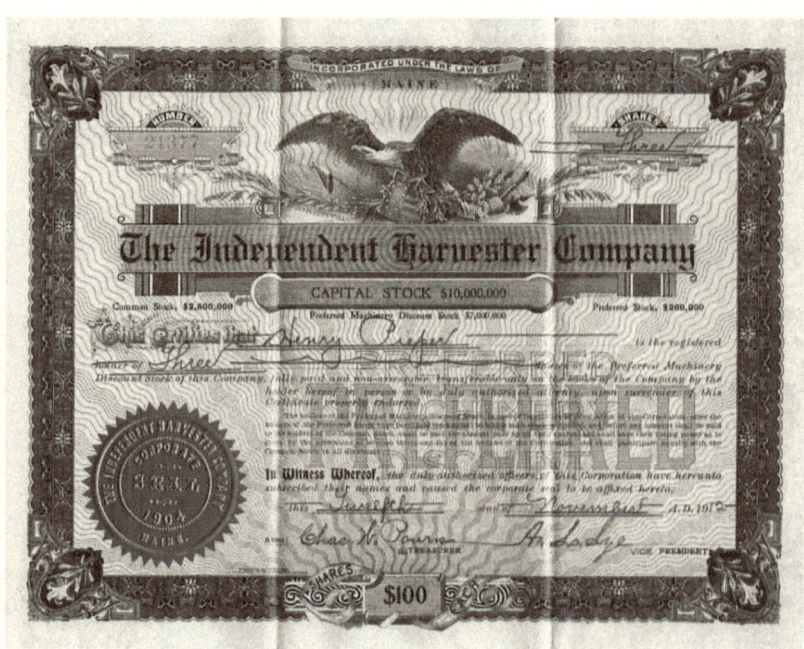

On June 15, 1909, the *Morris Herald* (Morris, Illinois) reported the suicide of Seneca, Illinois resident Fred Ehrman the
previous night, citing his concern about the loss of his hotel
through a failed business deal with the Independent Harvester
Company. Two weeks later *Farm Implements Magazine* announced that,

> *"..charges have been preferred against the Independ-
> ent Harvester Company of Plano, Ill., alleging fraudu-
> lent use of the mails, and that a government inquiry has
> been started to determine whether the methods of the
> company are legitimate or not. John Mahin, of Evans-
> ton, Ill., a postal inspector, recently visited Morris for
> the purpose of beginning the investigation. He called on
> Fred Ehrman of Seneca only to learn that Mr. Ehrman
> had committed suicide the previous night.... Whether
> Mr. Ehrman had preferred charges with the govern-
> ment or not is unknown, but he claimed to have lost
> about $5,000 in a deal with the harvester company, and
> this preyed heavily upon his mind and no doubt caused
> him to take his own life. Ehrman traded his hotel in*

Seneca for stock in the company, and claims he was offered a good position and other considerations which were not fulfilled.... While the name of the company is the Independent Harvester Company, they are not turning out any harvesting machinery as yet, but are making a manure spreader, a corn harvester and some other implements. But the principal industry at the present time appears to be the sale of stock."

The *Rural New Yorker* magazine picked up the story on July 10, adding, "For the benefit of those farmers who were inquiring recently about the wisdom of investing in the stocks of the Independent Harvester Company... It will be remembered that we advertised against the investment at the time."

After Ehrman's death there seems to have been a lull in the government's case against Independent Harvester. News of an investigation or progress in the case did not reappear for another two years.

Automobile manufacturer Henry Holsman of Chicago joined the Independent Harvester Company in April 1910 as a consulting engineer. Holsman had built his high-wheel cars in Chicago, beginning in 1903, but ran into financial trouble. He turned over his patent rights to Independent Harvester, who would manufacture the Holsman automobile in the Independent's Factory Number 2 building. The 1910 model K used a 2 cylinder 12 horsepower engine and a model M used a 4 cylinder 26 horsepower engine. New automobiles were coming on the market, but Holsman stuck with his basic design, described as essentially a motorized buggy. It retained "the tiller for steering, the cable drive and steel-rimmed wagon wheels long after the steering wheel, drive chains and mechanical drive shafts and pneumatic tires began appearing on other cars." (*Chicago Tribune*, June 21, 1998). The vehicles were advertised as suitable for farmers, doctors, and livery service. During their production years of 1910 and 1911 few were actually manufactured and Holsman's association with the Plano company was short lived. Confusion arose when the automobiles were referred to as I.H.C., for Independent Harvester Company, since the International Harvester Company also used the initials.

Independent H Commercial Cars.
Made by Independent Harvester Co., Plano, Ill.

UTILITY MODEL 22, $750.
Motor. 26 h. p., 4 cylinders, air cooled, jump spark ignition; loading space 96 x 31 in.; carriage type wheels, steel tires; starting handle can be used on either side of vehicle.

An example of Independent Harvester's questionable expenditures was the opening of the Independent Club in Plano in February 1911. The club had a clubhouse complete with a piano, billiard tables, card tables, a reading room, a library, kitchen, and a bathroom with tub and shower. The club opening was celebrated with an evening of presentations regarding the company's future, plus musical performances which included selections played by the Independent Harvester Band. The club's purpose was explained as a place for entertainment and education for company employees.

The company's slow start in actual manufacturing is revealed in the number of machines built in its early years. Between 1905 and 1911 the company built 51 corn binders. In 1909 they began building grain binders. The six built in 1909 were followed by 135 and 560 the next two years and 1,921 in 1912. Their mower production was higher, the firm building 4,065 mowers between 1910 and 1912.

By 1913, Independent Harvester manufactured fifteen kinds of machinery which they sold in fourteen states and had 27,000 stockholders. Department supervisors were:

Superintendent of Factory No. 3	Walter Scott Nichols
Assistant Superintendent of Factory No. 3	William Voorhees
Forge Shop Foreman	E. Frazell
Assembly Department Foreman	A.C. Gamill
Plow Fitting Department Foreman	C.H. Fleugel
Grinding and Polishing Foreman	Chas. Hull
Plow Tempering Foreman	Gust Anderson
Woodshop and Millwright Work Foreman	F.R. Nichols
Machine Shop Foreman	Arthur Benson
Pattern Shop and Experimental Department Foreman	C.H. Hunt
Paint Shop Foreman	Geo. A. Parker
Shipping Gang Foreman	Mr. Courtright

Front view of Independent grain binder, showing tongue truck and four-horse evener

Farmers who bought or considered buying stock in the company were interested in its stability and likelihood of future success. In 1912, a movement began among stockholders to get answers to their questions about the company's finances. A committee of Iowa stockholders in the Independent Harvester Company visited Plano in late 1912 to investigate the financial condition of the firm. The report of their investigation, made in conjunction with committees from four other states, was not encouraging. On the day William C. Thompson had agreed to meet with the twenty representatives of company stockholders, Thompson failed to appear. The 20 questioned company officers who were present, reporting their answers to be "dilatory, incomplete and evasive." Several days later, Thompson finally met with the committee, his responses no more enlightening than the officers'.

Farmers who owned $6,000,000 of Independent Harvester stock held a so-called "indignation meeting" in January 1913. They claimed that the corporation's stock was sold at inflated prices, as high as $25 above face value, and they demanded that company President William C. Thompson account for $350,000 in stock he allegedly held. He declined to do so. They alleged they were told that Independent Harvester Company had been organized to "buck the harvester trust off the map" but they had no voice in the management of the company which they believed had been grossly mismanaged. Thompson was the main focus of their complaints.

Twenty-six stockholders, each owning at least $5,000 of stock, toured the manufacturing facility in Plano in March, their purpose to investigate complaints about the company. The report of the 26, to be sent to all stockholders, stated that they were satisfied that the plant's planned expansion would enable the company to meet orders received and proclaimed their confidence in the company's management. The report did not quell discontent.

A committee representing 27,000 Independent Harvester stockholders, their indignation not appeased, filed suit in the United States District Court on May 2, 1913 alleging mismanagement of the company and demanding an accounting from the company's officers. They asserted Independent Harvester guaranteed big dividends as well as reduced prices on the company's products and claimed that the company's assets did not exceed $1,000,000. According to the suit, the Independent Harvester's management had "organized a stock selling campaign, and for four years or more have devoted all the time and energy of the officers and a large number of employees of the corporation to selling stock, have expended large sums in advertisements, employed sales agents for stock, commissions as high as thirty per cent of the sales... That stock sales were conducted for the sole purpose of paying salaries, expense accounts, and profits to the individuals in the management; that more than three million dollars' worth of stock has been sold during the past two years; that the management diverted the corporation's purpose from manufacturing machinery to the sale of stock; that it conspired to continue stock sales and neglect manufacturing; that the mismanagement and fraudulent misrepresentations of defendants so injured the reputation of the company and its products that advertisements were refused by farm and other journals..."

The complainants asked that Thompson be removed from office and the company be reorganized.

The following day, the *Chicago Examiner* interviewed Thompson who claimed that the company had been persecuted throughout its history. He contended that "the 'tentacles' of a 'vested interest' reached and tightened their grasp on commercial agencies, foundries and even to the editorial and advertising columns of newspapers all through the corn belt of the Northwest in an effort to starve its smaller rival through lack of opportunity to make and market its products."

The United States Postal Service investigation had reappeared in the news that spring, a local newspaper reporting that William C. Thompson was cooperating with the postal investigation and providing free access to the company's books. The newspaper

---

William Campbell Thompson
1864-1936

William Campbell Thompson came to Chicago from Londonderry, Ireland while still a child. In 1893 he married Marjorie Fraser, with whom he had one daughter, Marjory Fraser Thompson. Early in his career he worked in real estate and banking, mainly as a bonds dealer. Thompson formed the William C. Thompson Company in 1900, a dealer in municipal, railroad, and government bonds.

He joined the Kellogg Harvester Company in 1904 as Vice President and Treasurer. The company changed its name to the Independent Harvester Company in 1905. Thompson resigned from the Independent Harvester Company in 1913 as he and other company officials were under investigation for fraud. He worked in the insurance industry for the last ten years of his life. On March 26, 1936, while on his way home, William Campbell Thompson died on an elevated train in Chicago, reportedly of a heart ailment.

---

encouraged Plano residents to support the company and refrain from "unjust criticism."

On June 24, 1913, Stephen Gregory, attorney for the Independent Harvester Company stated that the officers and directors of the company would resign that day amid an investigation by the Department of Justice into mail fraud due to their methods of stock selling. The resignations and reorganization of the company would resolve the stockholders' lawsuit. The mail fraud investigation continued.

Those who resigned were William C. Thompson, President; Ning Eley, Secretary; and Directors A.K. Wentworth, Joseph Boda, A.L. Lye, W.W. Parish, Jr., and Robert McLeod.

Under the reorganization of June 1913 the officers of the corporation were William Deering Steward of Plano, President and Treasurer; Charles Emmett Jeter of Plano, Vice President; and Ning Eley of Chicago, Secretary. Eley resigned on July 15, 1913 and was replaced by Francis G. Hanchett of Plano.

Company directors as of July 1913 were William Deering Steward, Francis G. Hanchett, George S. Steward of Plano, F. L. Martin of Hutchinson, Kansas, E.M. Thebiay of Eagle Grove, Iowa, and Charles Emmett Jeter.

In May 1913 William Deering Steward had retired as mayor of Plano, having served as the progressive Democratic mayor of a staunchly Republican city for fourteen years. The popular mayor was being considered for the position of U. S. Comptroller of the Currency. His support in Illinois was substantial but he was not selected for the position. Instead, he was faced with the daunting challenge of putting the Independent Harvester Company aright.

The new officers and directors of the company were viewed with suspicion by the stockholders because of the mismanagement of the former officials. Their predecessors had exaggerated the company's assets and the new officials discovered there was little working capital. The new management took on the task of reducing expenses. William Deering Steward was paid $10,000 per year and Hanchett, as secretary and attorney, was paid $4,000.

William Deering Steward, was challenged in the mayoral race of 1911 by William Crimmin, head of Independent Harvester's stock sales department. Steward was a popular mayor whose winning margin was increased by a Crimmin campaign misstep. Crimmin had printed and distributed in Plano 1,000 pamphlets described by the *Kendall County News* as,

"...cowardly, unpolitical, unbusinesslike, and absolutely untrue except in so far as they referred to the business affairs of the Independent Harvester Co. which they had no more right to use to further their political ends than the editor of this paper. This was their undoing in the City of Plano. It doubled Deering Steward's majority and also the majority for licensing saloons in our city, and was also the means of humiliating many honest and sincere temperance workers and there was lots of them at work.

Was the Independent Harvester Co. trying to dictate and run the whole City and our public schools or was it just a few high salaried clerks at the office of this Co., who are attempting to do this very thing if their talking outside in the past month was an indication how they would put the town dry and meaning the Independent Harvester Co. could elect a mayor any time by touching a button?

It is impossible to make us believe that the Independent Harvester Co. board of directors and stockholders would sanction or tolerate any such thing and only condemn those who took part. There is not a business man in Plano, and we defy any one to contradict us that would not go out and fight night and day for the honor and credit of the Independent Harvester Co., but they also love their home and the City of Plano."

The reorganized company borrowed money to operate. According to the management, when questioned in regard to the federal investigation, the harvester business "does not permit a turnover of working capital but once in twelve months" so it was necessary, in order to do business of at least one million dollars a year, to have at least nine hundred thousand dollars working capital to pay expenses. Bankers familiar with the harvester industry agreed with the assessment.

The Independent Harvester Company's new management turned its attention to manufacturing. They sold almost no stock. A "through-the-dealers-only" policy was established. Selling directly to farmers and the problems it caused would be a thing of the past. At their annual meeting of the stockholders on Feb. 16, 1914, full voting power was granted to all preferred machinery discount stock.

The new management continued their efforts to make the enterprise a success with the hiring of G.H. Carver as General Manager in November 1914. Carver began his career in the harvester industry while still a teen, working for Gammon, Deering & Steward. He was the Superintendent of Plano Manufacturing Company for several years. Carver remained with Independent Harvester until the summer of 1917 when he returned to Chicago.

From July 1914 to July 1915 the Independent Harvester Company shipped 540 train cars of machinery from Plano, the largest one year number in the history of the railroad station in Plano, according to the Chicago, Burlington, and Quincy Railroad Company.

On April 27, 1915, the Federal Court issued indictments against Independent Harvester Company former officials and several of their former stock salesmen.

Indicted were: William Campbell Thompson, former President; Ning Eley, former Secretary and attorney; William Crimmin, former General Stock Sales Agent in charge of all stock salesmen; M.O Shoop, Herman H. Borchers, Henry Clay Borchers, Edward E. Preston, Ward K. Spain, and James F. Thompson (brother of William C. Thompson), stock salesmen.

The men were charged with using the United States mails in violation of the postal laws, specifically conspiracy to use the mails to defraud between 1907 and 1912. Four weeks later, the indictments were followed by charges against six additional former Independent Harvester sales agents, also for alleged use of the United States mail in a scheme to defraud.

The publication, *Farm, Stock, and Home* had published an "I-told-you-so" article seven months earlier, when the federal grand jury in Chicago had returned indictments against the officials. The magazine reminded its readers that it had exposed "the stock selling scheme of the Independent Harvester Co.," and had warned "its readers to let it alone." *Farm, Stock, and Home* claimed to have received many letters critical of the publication for "knocking a farmers' company that was going to knock out the trust"—the trust referring to International Harvester.

William Deering Steward

William Deering Steward
1872-1953

William Deering Steward, third son of Lewis and Mary Steward, was born in Plano, Illinois and lived there his entire life. For many years he was the chairman of Kendall County's Democratic Central Committee, having been chosen for that position shortly after his graduation from the Chicago College of Law. Steward served as mayor of Plano for fourteen years and was president of the First State Bank of Plano for fourteen years. He stepped down as bank president to lead the Independent Harvester Company when its officers and directors resigned under a cloud in 1913.

In 1934 Steward was appointed postmaster of the Plano post office, a position he held until 1949.

In 1916 the Independent Harvester Company advertised in publications such as *Farm Implement News* for individuals interested in becoming dealers for the company. The company promised to protect the dealers, giving them large enough territories to be profitable. With its 25,000 farmers as stockholders, the company advertised itself as "The Farmers' Company," which produced eighteen different farm implements.

The trial of William C. Thompson and other Independent Harvester former officials got underway in Federal Court in Chicago in January 1917. Arthur L. Sanborn was the judge. More than 100 alleged victims were in the Federal courtroom to testify.

After 37 days, Judge Sanborn took the case from the jury and declared a verdict in favor of the defendants. The court held that it had not been proven that the defendants intended to defraud purchasers of stock in the Independent Harvester Company.

From the time the new management took over in 1913 until the end of 1917 they had worked hard to improve the company's situation, increasing annual sales from less than $500,000 in 1913 to $1,150,000 by the end of 1917. After losing money each year from 1913 to 1916, the company finally showed a small profit in 1917. The new management's focus included the importance of the success of Independent Harvester as the largest employer in the town of Plano and the need to repair the reputation of the company and thus the town along with it.

In the second half of 1917 the bank creditors made it known that they did not see a successful future for the company under its current plan of operation, so Steward and the other company officials decided the sale of the company under the authority of the Court was in the best interests of the creditors and stockholders.

In December 1917 William Deering Steward and Phillip G. Clifford reported on the reorganization of Independent Harvester. Having been appointed receivers for the company, they explained that the receivership was not due to insolvency but to allow the company to obtain funds to successfully operate with the increased business it was experiencing. In a report they issued at that time, they noted $832,000 in stock notes receivable. These notes were taken by the previous management in payment for stock but it had proven almost impossible to collect on them.

Company by-laws made selling or mortgaging any of the company's property nearly impossible.

The stock sales scheme of the former management which promised buyers both discounts on products and a share in profits also provided large commissions on the sale of stock. These sizable commissions were a factor in the financial difficulties of the company.

On Dec. 1, 1917 the Independent Harvester Board of Directors issued a report to the company's stockholders. The report stated that, after a meeting of the entire board,

> "a resolution was unanimously adopted, reciting that after four years of sincere and arduous effort, under the most unprecedented and difficult conditions, to put the business of the Company on a solid financial basis, the Directors were satisfied that it was no longer possible under the present plan of organization of the company

to procure the necessary capital to carry on business of sufficient volume to show a profit, and that to attempt longer to continue business on the capital and credits available would be to involve the stockholders in further losses, with no prospect of making the business a success. The President was therefore directed to employ counsel and proceed as advised to procure the liquidation of the business and property of the Company."

The report went on to explain the difficulties Independent Harvester faced. In addition to the damage done by the prior directors, officers, and stock salesmen, the unusual conditions caused by World War I and lack of capital to develop the business were outlined. The costs of raw materials and transportation had doubled in the two years leading up to December 1917. As the report explained,

"Agricultural implements are sold on unusual and we believe unnecessary terms, and manufacturers producing them are obliged to use more money in their business than almost any other business would require. Purchasing material, paying for necessary labor and operating a business for practically one year before the returns for goods sold begin to materialize, require an unusual amount of cash in the business."

At that point in late 1917 the company had 24,150 stockholders. About 22,000 stockholders owned one to four shares each, seven stockholders owned 100 or more shares, with the rest owning five to ninety-nine shares each.

The officers were William Deering Steward, President and Treasurer; Charles Emmett Jeter, Vice President; George M. Dyke, Secretary; the directors were William Deering Steward, George M. Dyke, Charles Emmett Jeter, Charles Marsh Steward, Joseph L. McNabb, Grant Grinnell, J.F. Webb, and F.L. Martin.

Independent Harvester factory buildings

Independent 3 Horsepower Air-cooled Gasoline Engine

Rear View of Independent Harvester Grain Shocker in an Oat Field

Independent Harvester Company - Factory No. 2

Independent Harvester factory buildings

Harvesting with Independent Harvesters

The Independent Harvester Company's assets were advertised for sale in March 1918. Only one bid was received. The company was sold to the "Independent Harvester Company, Limited" for $604,506.21. With other expenses, the cost reached about $1,000,000. The stockholders—both preferred stockholders and common stockholders—would not recoup any of their investment unless they put up twenty percent of their original investment for preferred stock in the new company. Stockholders who did not accept the proposition would receive their proportion of the proceeds of the sale of the property less the expenses of the administration.

The purchasing syndicate included Lawrence Fitch, President (President of Western Malleables Co. Beaver Dam, Wisconsin and Treasurer of Globe Seamless Steel Tubes Company); A.J. Earling, Vice President (Chairman of the Board of Directors of Chicago, Milwaukee & St. Paul Railroad; Treasurer, Grant Fitch, Vice President of National Exchange Bank, Milwaukee, Wisconsin); Secretary, A.W. Wilbrand (Secretary of Western Malleables Company, Beaver Dam, Wisconsin).

On May 14, 1918, the Federal Trade Commission issued its report on the investigation of the Independent Harvester Company. According to the report, the investigation, directed by Senate Resolution 212 of the 65th Congress, looked into "the organization, conduct, financial status and methods of the Independent Harvester Company of Plano, Illinois" as well as the pending receivership and proposed reorganization of the company and the "disposition of the company's assets and stocks of implements and raw material on hand." The report attributed Independent Harvester Company's failure to mismanagement and insufficient capital.

Some minority stockholders protested the sale and reorganization of the company but the FTC investigation concluded that their allegations were without merit. The report stated that, under William Campbell Thompson's leadership, the Independent

Harvester Company was "operated mainly as a stock jobbing proposition. Stock salesmen were paid from 20 to 25 percent commission and the notes were discounted at about 10 percent. Undoubtedly, false representations were made. The affairs of the company were grossly mismanaged in that there was practically no bona fide effort to manufacture."

With regard to the period when William Deering Steward led the company, the FTC report noted, "Mr. Steward was not an operating man and his selection was made because he was a man of integrity and of prominence in Plano, Illinois, having been a former mayor of that city." The company, under Steward's leadership, was hindered by the bad "reputation created by the Thompson regime and the inability to obtain sufficient capital... The receivership is the result of the officers and directors of the Steward regime feeling that they had become sufficiently financially involved, and the feeling on the part of the banks that enough money had been loaned by them. Unable to borrow, the receivership naturally followed."

The report added that Steward had not previously been connected with the company, had not been a large stockholder, and had agreed reluctantly to assume leadership of the Independent Harvester Company. And although "There was undoubtedly errors of judgment on the part of Mr. Steward and his associates" there was "no such mismanagement of the corporate affairs as characterized the Thompson regime... the Steward regime made every effort to develop the company and keep the stockholders fully informed as to the progress of affairs."

In June 1920 the Moline Plow Company purchased the Plano factory and Independent Harvester liquidated its business. Although completed machines and material were not part of the sale, Moline Plow would provide repairs for Independent products.

The Moline Plow Company sold the harvester plant to the Plano Foundry Company and Federal-Huber Company in April

1924. The property consisted of Factories 1, 2, and 3, machinery, the office building, and all the residences and lots owned by Moline Plow.

Since 1924 most of the harvester factory buildings have been in use by various businesses. Farm machinery is no longer manufactured in Plano but the town will always be "The Birthplace of the Harvester."

BIBLIOGRAPHY

Bateman, Newton and Selby, Paul. *Historical Encyclopedia of Illinois and History of Kendall County.* Chicago: Munsell Publishing Company, 1914.

Commemorative portrait and biographical record of Kane and Kendall Counties, Ill. Chicago: Beers, Leggett & Co., 1888.

Ardrey, Robert L. *American Agricultural Implements; A Review of Invention and Development in the Agricultural Implement Industry of the United States.* Chicago, 1894.

Marsh, Charles W. *Recollections 1837–1910.* Chicago: Farm Implement News Company, 1910.

Steward, John F. *The Reaper, A History of the Efforts of Those Who Justly May be said to have made Bread Cheap.* New York: Greenberg, Publisher, Inc., 1931.

Hutchinson, William T. *Cyrus Hall McCormick.* New York: The Century Co., 1930-1935.

Deering Harvester Company. *Deering Harvester Official retrospective Exhibition, Development of Harvesting Machinery for the Paris Exposition of 1900: made by Deering Harvester Company, Chicago, U.S.A. Chicago:* R.R. Donnelley & Sons Co., 1900.

Hicks, E. W. *History of Kendall County, Illinois: from the earliest discoveries to the present time.* Aurora, Illinois: Knickerbocker & Hodder, Steam Printers and blank Book Makers, 1877.

Steward, Julian R. *Narrative by Lewis Steward Related to Me.*
   Plano: Typed document, about 1880.
Steward, Aurelius. *Early history of Kendall County, Illinois,*
   supplemented by J.F. And J.R. Steward. Plano: Typed document. About 1902.
United States. Federal Trade Commission. *Federal Trade
   Commission report to the Senate on the Independent Harvester Company.* Washington D.C.: The Commission,
   1918.

Magazine, Newspapers, and Periodicals:
   Chicago Tribune
   Farm Implement News
   Farm, Stock, and Home
   Harvester World
   Implement Age
   Indianapolis Star
   Inter-Ocean
   The Iron Age
   Kendall County News
   Kendall County Record
   Plano Mirror
   Sycamore True Republican

Advertising publications of Gammon & Deering Harvester
Company, Plano Manufacturing Company, and Independent
Harvester Company

Personal correspondence of John F. Steward and Charles W.
Marsh

Plano Steam Power ledger book. Plano, 1880-1883. Property of
Barbara Hoffman.

Holsman automobile photo courtesy of
https://holsmanautomobiles.com

INDEX

American Harvester Company, 74
Anderson, Gust, 94
Appleby, John F., 49-51
Bartlett, Samuel, 9
Benson, Arthur, 94
Boda, Joseph, 85, 98
Borchers, Henry Clay, 100
Borchers, Herman H., 100
Brown, Simmons, 7, 8
Chicago Harvester King, 42
Coddington, J.F., 46
Courtright, Mr., 94
Crimmin, William, 99, 100
Davenport Potato Cutter and
    Planter, 79
Deering Manufacturing Company,
    48, 60, 72
Deering, William, 3, 30, 34, 35, 39,
    40, 43, 46, 51, 52, 53, 55, 59, 63,
    74, 98, 99, 101, 103, 105, 109
Dixon, R.H., 45
Dyke, George M., 104
Earling, A.J., 108
Easter, John D., 16, 23
Ehrman, Fred, 1, 91
Eley, Ning, 85, 98, 100
Elward, John H., 30
Fields, E.C., 45
Fields, Elijah, 66

Fitch, Lawrence, 108
Fitler, Edwin H., 51
Fraser, Donald, 85
Frazell, E., 94
Furgeson, Hugh, 46
Gamill, A.C., 94
Gammon & Deering, 47
Gammon & Steward, 35
Gammon, Deering, & Steward, 39
Gammon, Elijah H., 16, 23, 25, 29,
    30, 34, 35, 37, 39, 41, 43, 45, 46,
    51, 55, 59, 63, 66, 68, 72, 76, 82
Gordon, John H., 39
Gregory, Stephen, 98
Grinnell, Grant, 104
Haiseldean, George, 46
Hanchett, Francis G., 98
Haskell, Barker & Aldridge, 6
Henning & Ross, 46
Henning, A.E., 65
Henning, Gilbert Denslow, 15
Henning, William T., 60
Hibbs, William, 85
Hinds, Albert, 9
Hollister, John F., 4, 5, 13, 20, 21,
    46, 58, 65, 66
Holsman automobile, 92, 112
Holsman, Henry, 92
Hunt, C.H., 94

113

Independent Harvester Company, 1-3, 85-110, 112

Independent Harvester Company, Limited, 108

Jenks, Daniel, 87

Jeter, Charles Emmett, 98, 104

Johnson, Christian L., 48

Jones, O.W., 76

Jones, W.H., 63, 65, 66, 72, 74, 76

Judd, Thomas, 6

Kellogg, E. M., 82, 83, 85

Kennedy, Herman N., 42, 46

Lewis Steward & Company, 31, 32, 33, 35

Low, Adams & French harvester, 30

Lye, A.L., 98

Marsh Brothers & Steward, 23

Marsh Harvester, 2, 6, 11, 13, 15, 16, 23, 24, 29, 31, 33, 35, 42, 43, 58

Marsh Harvester Manufacturing Company, 31

Marsh, Charles W., 6-17, 27, 28, 112

Marsh, Steward & Company, 25, 29, 30

Marsh, William Wallace, 6-18, 23, 24, 26, 28, 31

Martin, F.L., 104

McCormick Harvesting Machine Company, 24, 36, 39, 74

McCormick, Cyrus, xi

McLeod, Robert, 98

McNabb, Joseph L., 104

Moline Plow Company, 109

Morris, C.M., 79

Morris, J.D., 79

Nichols, F.R., 94

Nichols, Walter Scott, 94

Owen, W.W., 79

Parish, Jr., W.W., 98

Parker & Stone, 16

Plano Implement Company, 79, 80

Plano Manufacturing Company, 62, 63, 65-79, 82, 86, 100, 112

Plano Steam Power Company, 62, 63, 68

Powell, John Wesley, 31

Presher, Lott, 5

Preston, Edward E., 100

Rodgers, Israel, 47, 48

Sanborn, Arthur L., 102

Sanders, A.C., 79

Sandwich Manufacturing Company, 30

Sandwich, Illinois, 48, 52

Sears Manufacturing Works, 80, 82

Sears, Albert H., 55-58, 62, 63, 66, 80, 82, 85, 86

Sears, James M., 82

Spain, Ward K., 100

Steward & Marsh, 22

Steward, Amasa, 30, 31, 36

Steward, Charles Marsh, 104

Steward, George H., 24, 33

Steward, George S., 87, 98

Steward, John F., 5, 13, 16-18, 23, 29, 31, 32, 35, 36, 40, 51, 112

Steward, Julian R., 85

Steward, Lewis, 3, 12, 13, 15, 19, 24, 31, 32, 33, 35, 41, 43, 60, 62, 63, 68, 76, 112

Steward, Marcus, 4, 5, 6, 12

Steward, William Deering, 3, 98, 99, 101, 103, 104, 109

Thebiay, E.M., 98

Thompson, James F., 100

Thompson, William Campbell, 2, 82, 83, 85, 86, 95-98, 100, 102, 108, 109

Turpin, J.E., 79

Voorhees, William, 94

Walter A. Wood Mowing and Reaping Machine Company, 39

Warder & Mitchell, 23

Warrior Mower, 42, 45, 65, 71

Webb, J.F., 104

Wentworth, A.K., 98

Wilbrand, A.W., 108

Wood, L.B., 76

ABOUT THE AUTHOR

A Florida native, Jeanne Valentine has lived in Plano, Illinois since 1987. She co-authored (with Anne Sears and Kristy Lawrie Gravlin) the photo history book, *Images of America: Plano* (Arcadia Publishing) in 2012. Valentine began her career at Plano Community Library in 2002 where she is the moderator of the library writers group.

www.ingramcontent.com/pod-product-compliance
Lightning Source LLC
Chambersburg PA
CBHW022007170526
45157CB00003B/1186